THE POLITICAL ORIGINS OF INEQUALITY

THE POLITICAL ORIGINS OF INEQUALITY

Why a More Equal World Is Better for Us All

SIMON REID-HENRY

THE UNIVERSITY OF CHICAGO PRESS

Chicago and London

Simon Reid-Henry is associate professor in the Department of Geography at Queen Mary University of London and a senior researcher at the Peace Research Institute Oslo. He is the author of *The Cuban Cure: Reason and Resistance in Global Science,* also published by the University of Chicago Press.

The University of Chicago Press, Chicago 60637
The University of Chicago Press, Ltd., London

Printed in the United States of America

25 24 23 22 21 20 19 18 17 16   1 2 3 4 5

ISBN-13: 978-0-226-23679-7 (cloth)
ISBN-13: 978-0-226-23682-7 (e-book)
DOI: 10.7208/chicago/9780226236827.001.0001

Library of Congress Cataloging-in-Publication Data

Reid-Henry, Simon, author.
The political origins of inequality : why a more equal world is better for us all / Simon Reid-Henry.
pages cm
Includes bibliographical references.
ISBN 978-0-226-23679-7 (cloth : alkaline paper) — ISBN 978-0-226-23682-7 (ebook) 1. Equality—Economic aspects. 2. Economic geography. 3. Political geography. 4. Distribution (Economic theory)—Social aspects. I. Title.
HM821.R45 2015
330.1—dc23

2015012862

⊗ This paper meets the requirements of ANSI/NISO Z39.48-1992 (Permanence of Paper).

To set aside the sympathy we
extend to others beset by war
and murderous politics for a
reflection on how our privileges
are located on the same map
as their suffering . . . is a task
for which the painful, stirring
images supply only an initial
spark.

—Susan Sontag,
Regarding the Pain of Others

CONTENTS

INTRODUCTION

OCCUPY YOURSELVES! THE GLOBAL 1%

There is no wealth but life.
—John Ruskin, *Unto This Last*

We have reached a crossroads in our history. For all the achievements and riches of our time, the world has never been so unequal or more unjust. A century ago, at the time of the First World War, the richest 20% of the world's population earned eleven times more than the poorest 20%.[1] By the end of the twentieth century they earned seventy-four times as much.

Measured in terms of wealth, rather than income, the picture is even more extreme. Globally, the richest 1% now own nearly half of all the world's wealth. The poorest 50% of the world, by contrast—fully 3 billion people—own less than 1% of its wealth.[2]

There is growing awareness today of the consequences in rich countries of rising inequality: we know what it means to talk of the 1% there. But when it comes to the much greater gaps between rich and poor the *world* over, we confine ourselves still to talk of "global poverty". How often are we told that, if only we could see what life is like in a cramped slum in Dhaka or on some scrabble of land in rural Chad, we would be moved to help?

But the problem is not one of empathy. We are all familiar with the shape of a human body in hunger. The details, like glass paper, scarcely catch the imagination any more. It is not one of distance, either. A growing number of the wealthiest people in this world live in high-rise apartments that tower up and over the slums below—and they know only too well that before all the "beautiful forevers" will be lived a thousand impossible todays.*

The problem, rather, is one of perspective, of what we choose *not* to see. There is no shortage of books telling us "why nations fail" or what "the bottom billion" on this planet must do to succeed, no shortage of policy papers from the World Bank or the International Monetary Fund saying much the same. But we still have not properly confronted how the poverty and suffering of a great many are connected to the wealth and privilege of a few.†

We are slow to admit that the problem is one not of poverty traps at the bottom of the pyramid but of a great confinement of wealth at the top. It is a telling coincidence indeed that the past fifteen years, a period when global wealth more than doubled (from $117 trillion in 2000 to $263 trillion in 2014), have also been the age of the Millennium Development Goals (with their headline ambition of halving the proportion of people living on less than $1.25 a day).[3] Those goals end this year, in 2015, in many cases not having been met. Meanwhile, global wealth keeps on growing: by 8.3% from mid-2013 to mid-2014 alone.[4]

There is a politics to this, but it is all too often ignored in a debate which to date has preferred to focus on the economics of the problem alone: as if the long-run dynamics of capital and income could be separate to the political history in which they are set. The primary purpose of this book is to paint this wider political context back into the picture, since our problems stem less from market forces than from the failed policies behind them. If this is partly cause for despair, then it is also cause for hope: our present predicaments are more amenable to change than we are often encouraged to believe.

But acting on the politics of inequality requires first grasping the full scale of the problem before us. Few of the world's richest people

*Katherine Boo, *Behind the Beautiful Forevers: Life, Death and Hope in a Mumbai Slum* (London: Portobello Books, 2013).

†James A. Robinson and Daron Acemoglu, *Why Nations Fail: The Origins of Power, Prosperity and Poverty* (London: Profile Books, 2013); Paul Collier, *The Bottom Billion: Why the Poorest Countries Are Failing and What Can Be Done about It* (Oxford: Oxford University Press, 2008).

intentionally exploit the world's poor, it is obvious to note, and most of us are not personally responsible for the plight of distant strangers. But despite seven decades of international development, three decades of the Washington Concensus, and a decade and a half of Millennium Development Goals, the world today is more divided than ever among the haves, the have-nots, and—as President George W. Bush liked to quip over an after-dinner brandy—the have-mores. Class still rules our world, in short. But geography is now enlisted in its cause.[5]

This means that inequality is felt differently at different scales, even as those scales are increasingly connected.[6] In rich and poor countries alike, however, and perhaps above all between them, inequality is a product of ingrained norms of status and rights that disqualify the needs and claims of some relative to others. The making of those norms in modern society is the history of the struggle between forces that seek to privatise public gain and forces that seek to nurture strong societies. To understand inequality, in its fullness, requires grasping this first of all. The exclusion of the poor via migration controls internationally, for example, is of a piece with the exclusion of particular social groups, such as Latinos, within a city like Los Angeles: there is a common core to both problems that it is possible to isolate and understand.

There are an increasing number of reasons to try. The extent to which the rich and powerful are today able to influence rules and procedures in a country like America is in part a product of the wealth they have accumulated elsewhere in the world; the struggle to make ends meet experienced by ordinary folk in austerity-ridden Greece is of a piece with very probems that the poorest of this world have confronted all along. Accordingly, there is no singular "us" versus "them" around which to build a greater justice. No clear-cut boundary to the problem of inequality. There *are* ongoing structures of exclusion and marginalisation and the imperative to try to understand their operation.

Hanging above the UN Security Council chamber is a vast mural painted by the Norwegian artist Per Krohg. The mural depicts a world rebuilding itself, after two world wars, on the principles of democracy, human rights, and equality. Seventy years later, and like the phoenix that Krohg places at the heart of his picture, this book opens upon the scene of a world still divided. In rich and poor countries alike, inequalities are becoming an ever more visible feature of contemporary society. Yet the origins of inequality—which have to do with questions of distribution, status, and authority—are so diverse that the larger structure itself often seems impossibly hard to pin down. This structure is so deeply embedded in the world—it seems so natural—that often we simply do not notice it properly. In this we are like the two little fish happily swimming along who meet a bigger fish swimming the other way. The bigger fish says good morning and asks them how the water is. The two little fish swim on for a bit until one of them suddenly turns to the other and asks, "What's water?"[7] Today we all need to do better at finding out how the water is.

The incentive could hardly be greater. Inequality is the "fount and matrix", to borrow a phrase from the great economic historian Karl Polanyi, of a great many of our era's most pressing global problems, be it climate change, food insecurity, economic volatility, or the demographic crises of migration and population growth.* These problems will all resist even our best efforts to address them until we have the gap between poverty and privilege the world over more firmly in hand. Poor countries will not join us in ambitious carbon-cutting goals if they have no other means to sustain themselves.[8] The world elite's anxieties about migration and security will not go away if we cannot reduce the incentives for others to leave their own moribund nations or for economic actors to trade on terms that they may take hope or reason much to care from.

In difficult times we are tempted, if not encouraged, to look inwards, to lower our ambitions and shutter up our hopes. But global inequality is a challenge that we can meet only if we are prepared to do the opposite of that which conventional wisdom supposes: to look upwards and outwards, to think bigger not smaller, and to confront head-on the very heart of a problem which ails us all.

It is precisely here that our political imagination fails us. Our politics remains stuck in the past. Worse, it is stuck in the wrong past.

*Karl Polanyi, *The Great Transformation: The Political and Economic Origins of Our Time* (1944; Boston: Beacon Press, 2001), 3.

Instead of imagining new ways of meeting the changing needs of society, we take comfort from well-worn clichés. Two centuries after the Enlightenment, when questions about the rise and fall of nations were first raised in earnest, today's commentators continue to peddle an endless stream of "decline of the West, rise of the rest" narratives, sharing with us their misanthropic visions of a zero-sum world in which winner takes all (and the losses of everyone else are overlooked). We seem unable to devise the necessary paradigms for our own times.*

If we could stop approaching the problems of our current political and economic order as if they were a parlour game, we might find it easier to identify and address the real challenges before us. At the very least, we might see that it is not "others" who threaten us in the first place—be they immigrants from whom we are told to protect our jobs (and fear for them), or the threats of an unspecified terrorism that we give up our civil liberties to counter. We are threatened much more directly by the facts of an unequal world and the consequences of those facts left to fester.

Left unaddressed, and inequality cuts into people's health and education, and into the life chances of us all. It does this both within our own nations, at home, but also for others abroad. It erodes what is left of our national and local communities and prevents us from exploring more peaceful relations with those with whom we share our planet. And since we experience these effects of inequality socially, in addition to the toll it exacts upon us as individuals, we will ultimately pass them on to our children.

As recently as the roaring 1990s, it was common for people to believe that the gap between rich and poor simply didn't *matter* any more. Time and again it was said that a rising tide of wealth would float all other

*Niall Ferguson, *Colossus: The Rise and Fall of American Empire* (London: Penguin, 2009); Gideon Rachman, *Zero-Sum World: Politics, Power and Prosperity after the Crash* (New York: Atlantic Books, 2010); Dambisa Moyo, *Winner Takes All: China's Race for Resources and What It Means for Us* (London: Penguin, 2013); Ian Morris, *Why the West Rules—For Now: The Patterns of History and What They Reveal about the Future* (London: Profile Books, 2010).

boats. Since the onset of the 2007–2008 financial crisis, such confidence in the notion of trickle-down growth—the Kuznetsian belief that inequality now is necessary if we are to get to prosperity later—has rather dried up.* But without it, neither side of the political spectrum seems to have the answers any more.

Internationally speaking, many on the right have responded to the recent economic downturn by retreating once again to the prejudices of the past: insisting that the world's poor are poor through lack of ambition, lack of skill—the product of "cultures of poverty" and idle dependency still. No to "global welfare" screamed Britain's *Daily Mail* in 2012, as the Conservative-led British government dared to maintain the previous Labour government's commitment to increasing international aid contributions.[9]

But the left is short of answers too. The Occupy movement has drawn attention to the unfair share of wealth held by the top 1% of income earners in rich nations. Yet a good few of those protesting in New York's Zuccotti Park in 2011 were themselves a part of the *global* 1% (at the time, so was anyone with an average annual income of $45,000 or more).[10] This doesn't invalidate their argument, but it does reveal a characteristic short-sightedness. The Indignados in Spain, Syriza in Greece, even middle-class student protestors in Chile—all are right to draw attention to the unearned privilege of a domestic superrich in times of national austerity: average wealth in the United States is now 19% more than even its pre-2008 peak, and in 2010, no less than 93% of post–financial crisis recovery in gross domestic product in the United States went to the top 1% of income earners. We are living in the age of Gatsby again.[11]

Yet what all too often goes unremarked in this debate is the fact that rising inequality in the United States and other wealthy countries is not, and has never been, isolated from a larger story about the international economy and international politics, which connect us to people and places we will likely never even hear of. Against this we need to ask ourselves, honestly, where it is we have learned our tolerance for in-

*The economist Simon Kuznets famously argued in 1955 for the inverted-U curve: the idea that inequality will rise as an economy develops, but then fall again over time as it matures.

equality at large; we must ask where the power of banks and moneyed interests comes from and how, in different ways, it is wielded over all of us; we must ask ourselves why the international decisions purportedly taken in our name seem to benefit just the usual few, and we must ask whether this is unrelated to the level of interest we ourselves pay to the ballot box. For all the talk that inequality is driving us apart, the politics of rich and poor the world over are, as I want to show in this book, in fact more closely entwined than ever before. And this being the case, the manner in which that politics is to be managed is of the utmost relevance, not just in terms of socioeconomic justice but also in terms of the viability of democracy itself.

There is a *geography* as much as a politics to inequality, in short, and when it comes to addressing the root of the problem, we need to keep this very much in mind. This poses two particular challenges. The first challenge is implicit in the fact that, while we now have a healthy debate in the United States and Europe about the historical *return* of inequality to our nations, we are blind still to the issue of inequality at the global scale. And yet, internationally, of course, inequality never went away. That we choose to overlook this is in part because the discussion of national inequality in countries like Britain and America latches firmly onto wealth as the problem (to wit, we vent our fury at the bankers) while the discussion of global inequality continues to focus almost exclusively on the poor (insisting that they lack this or that attribute: democracy, property rights, or the necessary "work ethic" most usually).

Changing this requires not only establishing why some people are richer or poorer now than they were in the past—this has been Thomas Piketty's great contribution—but also recognising exactly how privilege and suffering are hardwired into the world.* We are accustomed now to hearing about the economics of inequality. But wealth and power are *structurally* embedded in *political* society at large. They are also embedded internationally. Yet our present debate shows almost no interest at all in the plight of those who are daily kept in outright poverty

*Thomas Piketty, *Capital in the Twenty-First Century* (Cambridge, MA: Belknap Press of Harvard University Press, 2014).

by the incessant chasing of wealth elsewhere. Those who do show an interest, by contrast, are often the very individuals and corporations seeking to play the absolute poverty of the one (Bangladeshi sweatshop labourers, for example) against the relative poverty of others (Western workers, usually).

But history suggests that there are more hopeful connections we might draw here: coalitions can be forged and compromises found between people of differing outlooks and needs; individuals can overcome their prejudices of other people and places; and—by changing the scale at which they conceived of intransigent problems—societies in the past did find the tools they needed to address the extent of inequality in their societies.

There are good reasons, then, to pay attention to the history of social and political inequality in different places, since it tells us not only about those policy regimes under which inequality grows but also about the impact of policies conceived in opposition to it. America's New Deal and Europe's Golden Age, the first debates over modern democracy in post-Napoleonic France, the Kanslergade and Saltsjöbaden class compromises in Scandinavia, the granting of full suffrage to women and the US civil rights movement—all these and more are object lessons that inequality and the injustices that sustain it need not be an inevitable trajectory anywhere, and that answers to difficult questions can usually, with a modicum of political acumen, be found.

It may well have been the widening gulf between rich and poor in the early twentieth century, for example, that delivered the world into the Great Depression, and then in turn gave fascism and Soviet communism grounds for making their historical advances. But it was the fallout from the Second World War that would ensure that most Western countries never again fell prey to the extremism of either side of the political spectrum (until recently, perhaps). And it was that atmosphere of postwar tolerance that led most of the more successful rich nations to oversee a period of state-regulated capitalism, thereby narrowing their levels of domestic inequality and increasing social inclusion.

In most countries it was the war itself that kick-started this period: it was war that brought society to its senses after the excesses of the Gilded Age, and war that levelled incomes as much as cities across the

West, clearing the way for the Golden Age that would follow. But surely we can found a more positive politics of change for our own times, one that does not leave everything to war or chance alone? At the very least, the history of those times and places where inequality has been reduced points us to the importance of expanding political agency rather than closing it off. History teaches us that in times like ours, we need more, not less, politics: but we need to learn this history first of all.

Yet if we are to avoid the temptations of cynicism in this task, we must equally avoid nostalgia: and here too geography proves a useful guide. There is certainly an art to recognising how deeply ingrained problems, intractable though they seem to us as individuals, can be solved when we decide to settle our differences collectively. And we would do well, I am suggesting, to rediscover this. But we need to do more than simply hark back to the Western past. For all its achievements, the Golden Age shone brightly for relatively few people on this planet. The historical narrowing of the gap between rich and poor in just a few fortunate countries in the West never carried over to the gap between rich and poor countries at large. And when we look back at the twentieth century, this surely counts among the least remarked of its many catastrophes. It is without doubt into yet another unmarked grave of that century that the achievements of those same more fortunate countries are now themselves being flung at the hands of the market neoliberalism and liberal democratic triumphalism that have monopolised Western politics since the end of the Cold War.

At the very heart of this book, then, is the claim that we need a new politics to sustain us into the twenty-first century. We need a new strategy of equality to get us there too. Inequality may be felt most strongly within any one national community, but the challenge of inequality is a *global* one. Meeting it must begin, as the French political historian Pierre Rosanvallon says, with "rethink[ing] the whole idea of equality itself."[12] But it must also involve, in the twenty-first century, rethinking the scale at which equality needs to be made. As we shall see, this implies a series of rather more practical concerns. Not least, it serves as a timely reminder for Europe and America, in particular, to practice a little humility, to look outside their own treasured histories for solutions to the problem that they, as much as any other, now face. We rail today at the injustice of paying off the debt incurred by banks that were bailed out during the credit crisis, but this is what the Third World debt burden has always been about—common people paying off debts incurred by corrupt elites. There are plenty more connections too, if we care to look.

The second challenge arises to the extent that we are able to make progress towards the first. It concerns the fact that the more general crisis of the Western welfare state that we have been experiencing of late is not itself unrelated to the ongoing injustice of uneven *global* development. The runaway increase in wealth that today accrues to the richest of this world is what prices something like health care out of reach for even the moderately well-to-do; but these inequalities of wealth that are undermining the Western welfare state are produced through an international economy that itself comes to rest on the backs of the global poor, denying them their due of global public goods as well. The situation, as it confronts us today, is therefore simple: either we push forward with social protection globally, or we will see it continue to fall apart in the global North as much as in the global South. On one level, it might be said that these connections are too indirect, the causality too diffuse. Inequality is natural, we are told time and again. Of course we should expect to see it everywhere we look. But such claims, which are usually made by those with a vested interest in the status quo, confuse non-equality with inequality. Non-equality is desirable. It means difference, diversity, plurality, variety. Inequality is a product of structural forms of injustice. It means marginalisation, discrimination, neglect. And a system which serves some more than others.

This, then, is the wider landscape of inequality about which this book is concerned. And it is for this reason that it makes sense, I think, to speak of the *political origins* of inequality. To speak of just origins would be quite wrong: it would suggest a singularity of cause and a primacy of effect that simply does not pertain. Inequality is not a thing that we can work upon directly, in the hope of making more or less of it; it is a product of the wider choices a given society makes about how to organise itself, and the constraints placed upon that society from within *and* without. That being said, it is not entirely elusive, either. And from amongst the observable pattern of what passes for modern politics today, there are some things that more obviously need to be changed than others.

To introduce the qualifying adjective "political" is important, then, because by doing so, we accept that the dynamic of inequality (and it is a dynamic) is part inherited, part resisted, part intended, and part simply unrecognised: we are in a mess, in short, because of some things we have knowingly done, but also because there are other things we felt we really didn't have much choice about (mainstream economics being both author and beneficiary of the latter). This book is not a search for *the* origins of inequality, therefore, but an effort to understand the politics *of* inequality as that politics is made *between* people and places.

This is not, to be sure, the first time these sorts of questions have been raised. For the British historian R. H. Tawney, inequality was perpetuated in the modern world by a sort of historical bait and switch. Society civilises over time, naturally enough, but as it renounces the overt celebration of princes in their privilege, the sources of inequality are not so much outlawed as folded into the new institutional backcloth of society. As he surveyed the nature and extent of inequality in Britain and America in 1931—a fateful time to be doing so—it was thus all too clear for Tawney that, while inequality might no longer have been the "religion" that it once was in these countries, it was nonetheless a venerable tradition still.*

Tawney's work brims with insights still today. And yet when asked, at the end of his life, to write a new introduction to his classic work, *Equality*, he confessed to being no longer up to the task. His mental faculties were as sharp as ever. The problem, he felt, was that the nature and causes of inequality had become only that more much diffuse again: ingrained across a rapidly changing society and politics. In the Britain of the then 1950s, the political origins of inequality were slipping further out of sight. In the thicket of modern society, "ancient inequalities had assumed subtler and more sophisticated forms", as the sociologist Richard Titmuss put it.[13] And chief among those sophisticated forms was the fact that Britain's problems were no longer confined to Britain alone.

Tawney's problem was the same one that had defeated Jean-Jacques Rousseau over a century before.† Rousseau never elaborated

*R. H. Tawney, *Equality* (1931; London: Unwin Books, 1964).

†Jean-Jacques Rousseau, *Discourse on the Origin and Basis of Inequality among Men* (1755).

his *Discourse on Inequality* into an intended magnum opus "Political Institutions" (he burned the manuscripts and settled instead for the not-inconsiderable *The Social Contract*), in part because of his growing awareness "of the difficulty of his original undertaking". That difficulty, as the British political philosopher G. H. Cole wrote, was precisely that "he would have needed to deal not only with the nature of the social bond in any particular society, but also with the problems *of the relations between one society and another*, and with the whole question of moral obligations as arising for mankind as a whole".[14]

We are duly warned, then. And yet if Tawney got further than Rousseau (thanks chiefly to the invention, in the long century between them, of modern statistics), we have at least grounds, given the information at *our* fingertips, to speculate about some of those more diffuse origins as they concern *societies*, in the plural, today. For all the difficulties of this task, it is in any case imperative that we try. The best of our theories about inequality being based on a model of geographically closed societies, we are without the tools we need to make sense of our current, global inequality.

Now, it is true, as Thomas Piketty has argued, that it is the national scale that in many senses still matters. As Piketty points out, for all that many in the Western world today fret about the "rise" of emerging economies like China, the global future may well be one in which the poorer countries indeed catch up with the level of development of the wealthier ones. But in that longer term, their societies are also likely to have become more unequal. So touché. Perhaps. But there are two problems with this. The first is a moral one: must we really wait that long, knowing that we will merely have ended up with a more unequal world than our own, albeit a more "evenly uneven" one? The second is a theoretical one—why should we assume that their future will be the same as our past? This is the classic error of the diffusionist Eurocentrism that dominated the modernisation theories of mid-twentieth-century economies.

As against this we might expect, first of all, that other societies can and will draw upon different norms and institutions to manage their future economic relations from those we have thus far contrived to imagine for ourselves. Second, we should anticipate that we will ourselves want to take part in this process—that we may indeed stand to benefit from doing so. Both insights are in any case central to the discussion in this book. And it is perhaps here that Tawney, the twentieth-century historian, has something to teach Piketty, the twenty-first-century economist. For to properly understand inequality, Tawney argued, it was necessary to look not just at the dynamics and the composition of capital, but at the "the particular forms of inequality which are general and acceptable and the particular arrangement of classes to which they are accustomed".[15]

Ours being a global world, there is nothing for it today but to consider those forms in international perspective. Development is dead. The Western welfare state is in crisis. The challenge of inequality unites them both. If we cannot find reason to care what happens to distant others, then we will not find the means to help those who, through the split and the press of inequality, become strangers to us at home as well. And not finding reason to help them, we will be left to confront alone a range of modern problems whose solution we can find only together. This is not a challenge for the West alone. Much of the initiative must come from elsewhere. But the response does need to begin with the West, with a re-opening of the rich world, and of those who move within its circles of privilege.

This, then, is our starting point. World poverty and the inequality from which it stems need addressing today more than ever: they are, as we shall see, at the heart of so much else that ails us. But this will not come about through top-down policies of international aid and development (any more than will addressing the myriad associated problems of environmental degradation, economic instability, and ethnic strife). Nor will it come about if we merely button down into our own national pods and fret about inequality and the crisis of the Western welfare state, as if they were domestic problems alone. What is required, rather, is to establish the grounds for greater political and social inclusion at the national and international levels and to commit ourselves to more joined-up forms of democracy to ensure this. We must find just exactly where the twain can meet.

The message of this book is a simple one, therefore: "we" cannot end global poverty as individuals, but our societies, if we reinvigorate them politically, can reduce the inequality from which it stems. We can remove the venom from the bite and avoid the inevitable social backlash, which sooner or later will affect us all, even in our own communities back home. We can do all this without the enforced humility of war, as we shall see, by expanding on the lessons of two centuries of social policy and refitting them to meet our future, international needs. And since it is the underlying structural inequalities and the political forms determining them which together are the primary cause of our present discontents, it is with these that we are best advised to begin.

For if we can show (and we can) that the wealthy presently enjoy many of their privileges at the cost of others' continuing poverty, and if we can show (and we can) that neither rich nor poor are flourishing as they might, then we have the grounds for asking: Might not we all do better if those of us with a choice about such things found ways to include rather than exclude the poor, to engage with difference rather than seeking to manage it? What if, instead of projecting our fears onto others and accepting a world in which poverty and wealth grow ever farther apart, we sought to bring back in those we have excluded? What would *that* world look like?

This book is an exercise in finding out.

1

THE POLITICAL FORMS OF INEQUALITY

Man is born free; and everywhere he is in chains.
— Jean-Jacques Rousseau, *The Social Contract*

From the crisis in the eurozone to the US Congress shutting down, from food riots in Africa to Bangladesh's Rana Plaza caving in, our societies are more fractured and divided than ever before. And yet the problem is not that the world is falling apart. To the contrary, worlds we once thought separate are colliding. It is our ability to act in concert that is falling apart. It is this that needs restoring most of all.

We are not helped in this by the fact that our present language for making sense of the world has for some time now been corralled into a debate about economic globalisation on the one hand and development assistance on the other. Both vocabularies are rather less helpful than we are usually led to believe.

GLOBALISATION AND DEVELOPMENT

That our lives are fundamentally shaped in relation to a global economy is hardly news any more. But the more the world is subject to "global forces", the less it seems that we know what to do about them. And not knowing where to begin, we have allowed our politicians and our captains of industry to locate the causes of the problem elsewhere. Yet this denies the fact that today's global economy was itself built on a rejection of the earlier, Keynesian international order, and at a quite specifiable moment, the 1970s, when countries like the United States, the United Kingdom, and France all made the conscious choice to uproot the earlier unspoken commitment to regulating economic affairs. That earlier commitment had been put in place precisely to avoid the market instability that had led to the Great Depression. But in the 1970s, the rich countries believed, and they were right, that their nations stood to gain more from the market volatility that would result if the safety catches were once again removed.

The modern era of globalisation was inaugurated on distinctly uneven terms, then. But our problems today are about more than just the chickens of imprudent wealth creation coming home to roost. And holding globalisation itself responsible for what is done in its name all too often diverts attention from the more specifiable policies and practices we should be focusing on.

Too many of those protesting around the world today, for example, have too little to say about the way that emerging powers like India and Brazil are redrawing the lines between the winners and losers of globalisation. They have too little to say about the fact that many of the world's poor are now found *within* the borders of those states, where their own countries' corporations exploit them just as effectively as the Americans or Europeans ever did. Indeed, no sooner do countries like India, China, and Russia gain the attention of international investors than do questions of poverty and injustice invariably seem to disappear from the agenda altogether: a fact for which those countries' own governments are usually only too grateful.

Globalisation is today routinely presented as if its myriad economic and cultural exchanges add up to some larger, visceral force of nature: an unstoppable slither of fibre-optic cables and digital flows of money. In this reading, world poverty is little more than the patchy earth this beast chooses to pass by on. We have become so used to hearing about the "forces of globalisation" that we no longer bother to question what those forces actually are or whether they might be organised more fairly.

Perhaps this is not surprising. Given the tepid wash of compacts and summitry that passes for international politics today, it is indeed tempting to assume that democratic politics—reasoned political debate and argument, for those no longer sure—has nothing to do or say about the state of the world. But in truth there is much that politics—even national and local politics—*should* have to say about what globalisation is and does. We just need to start posing the right questions—or empower the right people to ask them for us.

The difficulty of asking the right questions, but also the reluctance to do so, is one reason the field of international development has, for more than half a century, served as a proxy for international politics. Development purports to include the global poor, but it has tended to so in ways that incorporate them as victims rather than as citizens, much less as equal ones. In rich countries today it sometimes seems as if there are few things for which an individual human right has not somewhere been enshrined. Yet when it comes to the global poor, we immediately start talking in aggregate: the struggles and the triumphs of individual selves get buried beneath our cloying talk of those who live on less than a dollar a day (now $1.25, of course).*

Such talk may be well intentioned, but it is also a part of the problem. It is politically easier to talk about increasing the proportion of the world's people who earn more than this magic figure than it is to point out that, for example, in 2012 the annual *increase* in the combined wealth of the richest one hundred billionaires alone was enough to lift everybody in the world above this line. It is politically easier to collect refugees together in camps that we then supply with blankets than it is to acknowledge that their problems are also our own.

This is one reason "emergency" has for some time now been just another word for "business as usual" for most of the rich-country-controlled organisations that ultimately do little more than simply manage the expectations of the world's poor. International organisations like the World Bank and the International Monetary Fund may be free of the Realpolitik and ideological self-interest that determined the international aid agenda during the Cold War. But in place of the overtly politicised aid of before, there is now almost no desire to set the problem of global injustice within any overt discussion of politics at all.

Meanwhile corporations, philanthropists, and unelected non-governmental organisations proceed to fill the gap with a constant

*There is, as it happens, a "right to development" too, put forward by Kéba Mbaye and adopted by UN declaration in 1986. But who remembers or very much cares about this?

deluge of poverty reduction "innovations". Too infrequently do those in positions of authority stop to reflect that this is not how today's rich countries developed, that we in fact did what we usually insist today's poorer countries *don't* do (such as protecting national industries behind tariff barriers). But it is not enough to cheer on the poor from poverty to power either, since poverty is a *product* of power. More than that, poverty *is* power—a form of power that some wield over others.[1] What the development industry has yet to fully come to terms with is the simple fact that it is *our* development that often stands in the way of *their* development.

There is a lot of talk within international organisations, for example, involving buzzwords like "protection" and "resilience". But it isn't always clear to whose better ends such words are being put, or at times what they even mean.[2] Of course, many of those working in these organisations would be the first to say so—but then they need to push back at what they are being told to do. What is the vision of Britain's minister for international development, we might ask? According to the aid expert Jonathan Glennie, "No one asks and no one cares; her job is to manage an aid portfolio apolitically."[3]

Glennie is right. Western governments and NGOs, along with many of the poorer governments they preach to, have increasingly come to favour the sweet pill of technocratic solutions focused on managing the symptoms (spinning the results when needed) to tackling the underlying problems head-on. They have learned to avoid the latter like the plague, in fact, because taking on those problems would demand that we make transparent the political choices inherent in today's uneven world order and have *those* be subject to public scrutiny. In truth this has long been "the challenge of world poverty", as the Swedish development economist and Nobel Prize winner Gunnar Myrdal put it nearly half a century ago.* Yet it jars with almost everything we are told about global poverty today.

So too do the facts of uneven global development. Stock-in-trade images of a bottom-billion world of landlocked nations mired in poverty

*Gunnar Myrdal, *The Challenge of World Poverty: A World Anti-Poverty Program in Outline* (New York: Pantheon Books, 1970).

traps or floundering at the foot of a development ladder are confounded by the fact that a great many people have been lifted out of poverty in recent decades: almost 1 billion between 1990 and 2010. All this is to be praised (though we should recall that it was China, not the Millennium Development Goals, which did most to contribute to the reduction).[4] But it remains the case that after "lifting" out of poverty, these people simply join a great many more whose problems are only marginally less acute. The structural causes of impoverishment and uneven development, meanwhile, which is to say the world's presently highly unfair ways of wealth, not only remain intact but go on largely unspoken of as well. There was little talk within the context of the MDGs about the no less remarkable fact that the global 1% saw its collective real income rise by 60% during the same period.[5]

Many experts insist on telling us still what the poor need to do to get richer, to become a little more like us: they speak accordingly not just of development ladders, but of "aspirant" and "climber" classes. Yet it seems that the poor are not found any more quite where these experts are looking for them. Their problems are certainly not what we are told to imagine. And it is far from clear, at the end of the day, that how to become like us is what keeps them up at night. Take a step back and it is in fact the geography and not the overall burden of poverty that is changing. Moreover, that geography is changing in such a way as to *increase* the future significance of inequality—by shifting it to middle-income countries, for example.[6]

All these changes have put the traditional development arguments of both the left and the right out of joint. On the right, calls to end aid altogether are increasingly common, especially now that austerity bites in the rich countries too, and aid seems like a luxury of the past. It may be true that poorly executed aid programmes become a crutch for local, unproductive elites, but cutting the aid lifeline altogether has no more humanity about it than when the British did this to their own poor in 1834 (throwing out the existing aid-in-wages system overnight, thereby "freeing" people to take responsibility for themselves).[7] On the left, it is more common to single out Western corporations and the "shock doctors" of free-market economics.

But both of those are merely acting in line with what international rules allow them to do. And those rules are set either by governments, which are voted for by people like us, or by companies, which respond to what their shareholders (including your pension plan and mine) demand. It is not just capital that matters, then, but the social structures and norms that justify and enable the accumulation of it. At the very beginning of the capitalist age, the division of economies into national units engaged in a zero-sum struggle was seen by Adam Smith and his contemporaries as a basic political root of inequality. Today we should look to the division of liberty itself into rights that can be "profitably" met by the market and freedoms which cannot.

Yet if the twin vocabularies of globalisation and development prevent us from seeing this, it is worth noting that such rhetoric is more recent than we sometimes recall. In the post-communist triumphalism of the 1990s—the first of two decades "consumed by locusts", as the historian Tony Judt put it—we ceded far too much control in the West to whomever and whatever seemed to promise us the greatest growth in our gross domestic product.[8] In the second of those decades—the one that began on the morning of September 11, 2001—we spent most of our time simply avoiding the consequences of our earlier naïveté, blaming poverty and the poor for the chaotic and disordered places we had ourselves created but now feared the next terrorist threat might come from.

But poverty is not simply the product of a "lack" of development or sufficient quantities of trade or materials to trade with. It is the product of uneven (and in many cases unfair) rules and the ability of some societies to manage such scant regulations as currently exist to their own short-term benefit. It is a problem of past political choices that have built up over time and the leverage this grants some people over others. It is central to the organisation of capitalism as a system and so central to the production of wealth.

Global wealth is likewise not just a measure of the value of a few people's "holdings" so much as a claim of entitlement to the running of the system itself. There is no reason to expect the geography of poverty and wealth to remain the same over time: just think of all the factories now lying deserted in Philadelphia and the pits that have closed in Yorkshire or the Rhine, as against the money that pours in to found construction booms in places like Dubai. But there is every reason to expect that when wealth is left to its own devices, the basic structure of inequality is destined to be perpetuated.

It is abundantly clear today, then, that the whole language of globalisation and development has been left behind by the facts of an unequal world, and left behind by the times. So too, unfortunately, have the slum dwellers of Caracas and Mumbai and São Paulo, while their compatriots drive, helicopter, and even funiculaire right past them on the way to work. What we need today are fewer platitudes about the benefits of a flat-world economy, a little less insistence that we can end global poverty in our lifetime, and a bit more recognition of the actual political and historical ways in which the lives of rich and poor are enmeshed.

There is a growing awareness of this today. But there is nothing like a clear enough appreciation of why addressing global inequality matters, either for ending the worst of the poverty that blights our world or for safeguarding the future of those of us who are, for now at least, somewhat better off.

ECONOMIC GROWTH AND POLITICAL INEQUALITY

Exactly why this matters begins to become clear when we look at the history of national GDP growth around the world over the past two centuries. One of the first things that immediately becomes clear is that the problem of uneven development is not primarily a problem of poverty in the first instance. It is a problem that began with the enrichment of some parts of the globe at the expense of others under colonialism.[9]

The "primitive innocence" that early modern Europe projected onto the cultures it encountered at the fringes of its ornate yet tentative maps of discovery was always a backhanded compliment. It merely served to confirm Western "modernity" in its sophistication. Meanwhile, Europe worked hard to create and manage markets among and between those parts of the world it controlled. In order to accumulate the wealth it needed to prove its own superiority, the British traded opium from Patna for tea leaves from China to ensure that both of those economies kept producing the tribute to which imperial London was growing accustomed.

This had little to do with "free trade". In India the British East India Company forced workers into the production of opium (actively undermining a previously thriving, transoceanic textile industry). In China gunboats were used to complete the sale there (all in the name of British shareholder value). Similar tactics were applied right across the colonial world. Settler populations in the United States and New Zealand received state support to corral, reserve, or otherwise prise local indigenous populations off their land. In southern Africa a nascent, European-financed mining industry did the same to local populations there, this time through the levers of colonial-state policies of war and taxation.[10]

The fundamental injustices of the colonial world order were then sustained into the twentieth century by the greater say that the wealthier countries' had in the legal, political, and economic running of the world. The dynamic of inter-imperial competition and a near-total breakdown of the international economic system eventually drove those wealthy nations into two world wars. The efforts of the United States to "resolve" the problems of a European colonial order soon led it to assert its own form of economic hegemony against this. But the same ability to determine the international rules and so to lock their own advantages into the emergent global system remained, and this

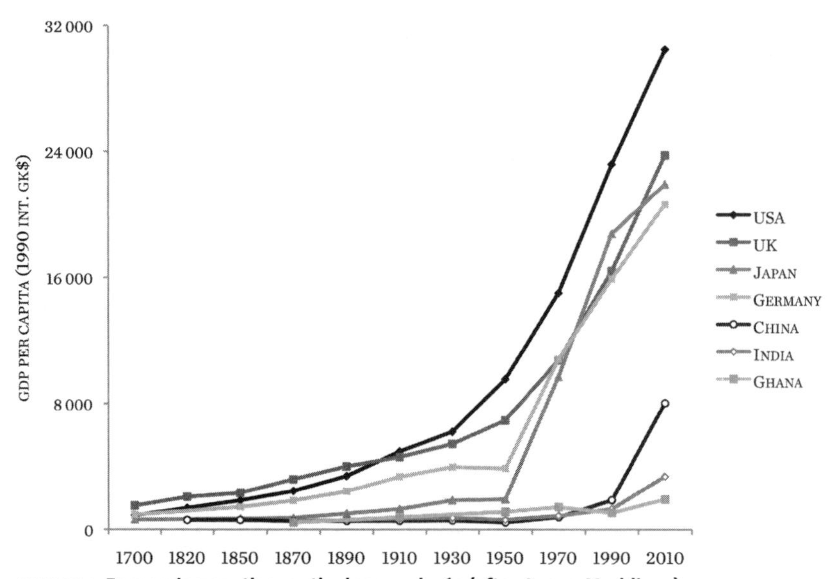

FIGURE 1. Economic growth over the longue durée (after Angus Maddison).

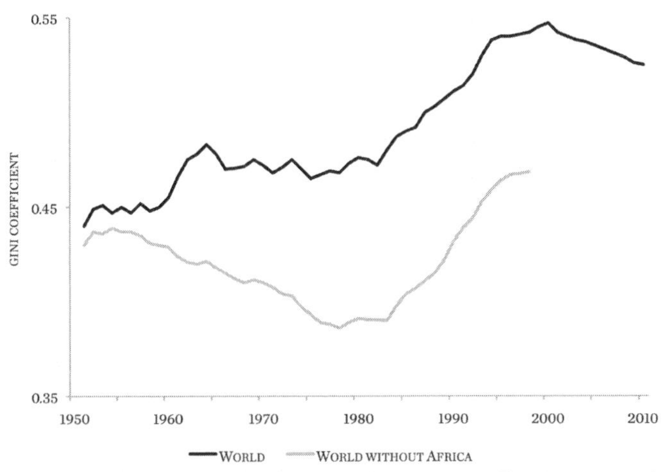

FIGURE 2. Historic inequality (non-population adjusted) in the twentieth century (after Branko Milanovic).

enabled the wealthier nations to again pull away from the world's poor after those wars, as the figures for their GDP per capita growth make clear. Led by America, and in part financed by the Marshall Plan, European nations and Japan powered rapidly ahead after 1945. Colonialism was on the way out, but the ghost of a world divided lived on (albeit, outside of Africa it was becoming more equal until the 1980s).

The modern-day market system that these earlier developments bequeathed is experiencing one of its periodic breakdowns again today. But, as ever, there are those who would deny it, the better to get things going again. Pick up the *Financial Times*, and amidst the despairing column inches on the euro crisis, you will also read about an Africa of "minigarchs" and "Afropolitans", a continent attracting investment from China and Brazil, one consuming Indian pharmaceuticals and talking on mobile phones made in South Korea. This picture is at one with the discourse of "emerging markets" so beloved of those who spend their Februaries in the Swiss mountain village of Davos (or in Africa, which now has its own economic forum, held each year in the local cantons of five-star hotels).

And yet for all that Africa *is* growing, it is doing so unevenly. Income inequality in sub-Saharan Africa is the highest by region in the world, and that inequality further encompasses considerable variation between countries.[11] In the 1990s, the Central African Republic was almost twice as unequal in terms of income as Ghana, whose own level of inequality was in about the same league as the United States. In the first place, then, Africa's poor are marginalised relative to the wealthy in their own countries. But they have also found themselves to be increasingly worse off relative to other regions of the world. In 1960 African nations accounted for just one-tenth of the world's extremely poor; by 1998 they accounted for two-thirds, a fact that makes them vulnerable to more predatory forms of foreign investment.[12] The image of Africa as a continent emergent thus presents us with no less partial a picture than the image of its people lying starving and neglected.

Elsewhere it is true that from the early 1980s, a handful of formerly middle-ranking countries—places like India and Brazil, and of course China, in its own exceptional way—leapt forward, with growth rates of up to 5%. Today, Brazil's GDP is approaching $2.5 trillion, and the country is nearing a permanent seat on the UN Security Council. The gap between America and Africa may remain such that only the very wealthiest of Ivoirians can compare themselves to even the poorest Americans in terms of income, but around half of all Brazilians can. And yet it is in countries like Brazil that resentment is often highest—as the 2013 riots in São Paulo over such seemingly trivial issues as bus fares indicate. Why riot? The reasons are not all that hard to divine: if the solution to poverty is simply to get rich, then this is of little help to the vast majority who remain poor.

For all that the gap between countries still matters, therefore, it is inequality within nations that is growing most rapidly today. This is the issue that has so preoccupied public debate in the Western world of late. But if within-country inequality is a serious problem in rich countries, then it is worth being aware of just how devastating it can be in poorer countries, where in many cases a child born into the poorest 10% of the population is today thirty-five times worse off than a child born into the richest 10%.[13]

In their 2009 book, *The Spirit Level*, Kate Wilkinson and Richard Pickett demonstrated the correlation between higher levels of income inequality and a range of social ills: everything from higher crime rates to greater levels of obesity.* As the figure here demonstrates, by plotting some of these same social factors against a standard measure of inequality for poorer countries as well (i.e., using a sample that goes beyond the rich OECD nations that were the focus of their work), a not-dissimilar pattern emerges. In countries where inequality is higher, homicide and infant mortality rates are both higher, while levels of educational attainment and health indicators both score lower, regardless of whether the country itself is rich or poor to begin with. But of course these effects are felt that much more acutely if the country also happens to be poor.

As a poor region there is no doubt that a higher level of economic growth would be desirable in Central America, for example. But there can be no serious talk of greater economic prosperity in a country like Honduras until the region as a whole can solve the problem of its displaced persons, who at present flee the cities and ride the trains up to the US border, in the hope of crossing over packed into the back of a coyote's truck. They will keep doing this, however, until cities like Tegucigalpa and San Pedro Sula, both of which have murder rates higher than that of Kabul, become safer places to live. And *that* will not happen until the more general gap in incomes and life chances between the inhabitants of those cities, and between those inhabitants themselves and the citizens of the southern United States included, can itself be reduced.

Just as in the rich world, then, the real problem that is tearing apart the social fabric in countries like Honduras is not poverty per se, but the effects of poverty as they are felt within a wider structure of inequality. And those effects are increasing not because the poor are getting substantively poorer but because the rich are pulling away: a process that is both abetted by local states desperate for any sort of growth at all and driven by wealthier international interests who are greedy for more. Corruption is an important part of this. Poverty matters too. But neither is the whole story by a long way.

*Kate Wilkinson and Richard Pickett, *The Spirit Level: Why More Equal Societies Almost Always Do Better* (London: Allen Lane, 2009).

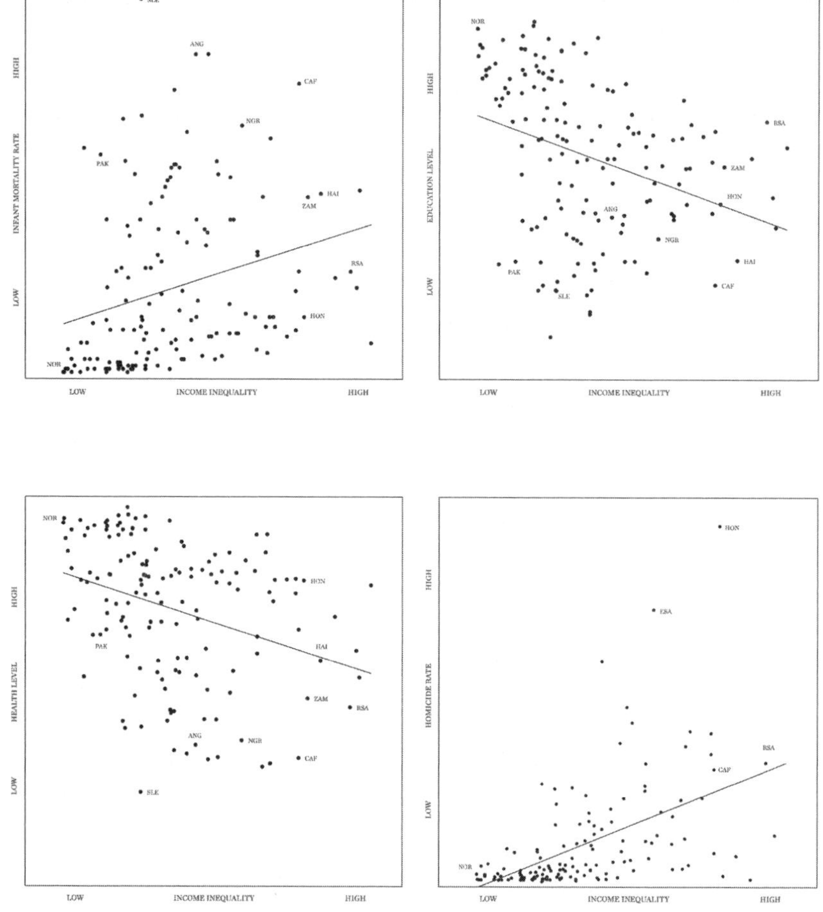

FIGURE 3. Infant mortality and homicide rates are higher in more unequal countries; educational attainment and health levels are lower.

In Asia, for example, where aggregate growth is frequently held to be the primary driver behind that region's growing affluence, had it not been for the fact that inequality has also been rising in recent years, as many as 240 million *more* people might have been lifted out of poverty, and all with not so much as a Millennium Development Goal in sight.[14] Indeed, while Asia is trumpeted as a success story in terms of economic growth, eleven of its twenty-eight economies (accounting for 82% of the continent's population) also saw rising inequality over the past two decades.[15] There is increasing evidence to suggest that when inequality reaches a certain level, it actually begins to undermine economic growth: a recent paper by the Organisation for Economic Co-operation and Development suggests the effect may even be "sizeable".[16] If this were so, then it would indeed be little wonder that poorer countries have thus far failed to catch up to the rich world.

Inequality is complex, however, and the picture varies depending on where we are looking and how we are trying to measure it. I have thus far talked primarily of inequality between nations and within nations (which are both problems needing urgently addressing). But a yet more damning picture of global inequality emerges when we look at the world as a single entity and compare ourselves not nation to nation, or even as weighted national populations, but person to person the world over as we each stand in our own skin upon it. When global person-to-person inequality is measured by the Gini index, for example, where 0 indicates perfect equality (we all earn the same amount) and 1 indicates perfect inequality (Bill Gates receives absolutely all income in the world, the rest of us work for free), the fact that such inequality has for some time hovered at around 0.70 would beggar belief if it didn't correspond so well to the world we see around us. After all, Britain's Gini index of 0.38 is high enough to give us everything from Roman Abramovich to the Aylesbury Estate, while El Salvador's level of 0.42 in 2011 contributed to the second-highest national homicide rate in the world (seventy-seven per hundred thousand, well above that even of America's most violent *cities* in the same year, Detroit and New Orleans, with rates of forty-five and forty-six per hundred thousand, respectively).[17]

The debate on inequality frequently gets caught up in a discussion about whether inequality is rising or falling, because the answer varies depending on what precisely we are measuring and how. The one thing we can really say with any confidence, however, is that whatever else it is doing, the level of inequality in our world is not going anywhere very fast at all, and it remains, all the while, stratospherically high. Taken as a whole, the real problem, then, is that inequality attaches itself, barnacle-like, to whichever social vessel matters in a given situation

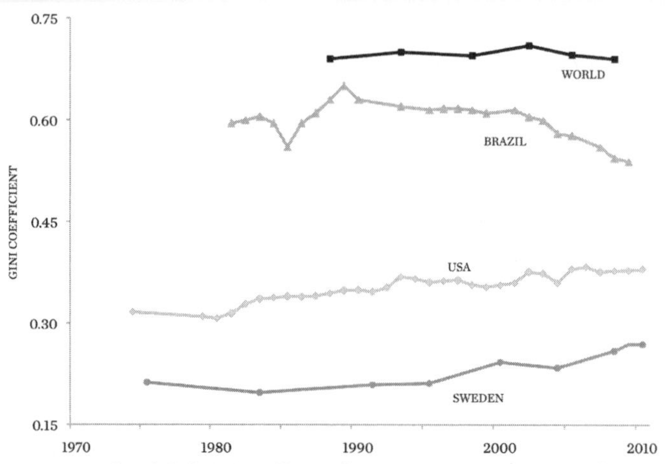

FIGURE 4. The global Gini coefficient (person-to-person inequality) compared to the Gini index of select countries (after Branko Milanovic, with some figures updated).

and, equally barnacle-like, remains in place until somebody comes along to prise it loose.

Given what has gone into making our world, this is why inequality globally has a rather particular geography, and it is why that geography is, as former World Bank economist Branko Milanovic points out, a strongly Western one, even today. The global 1% overwhelmingly reside in the West still (they are mainly wealthy Americans, Europeans, and Japanese, with a few of the very richest from other countries thrown in). If this group were to vote, it would vote in America and likely vote Republican (of the 60 million persons who make up the global 1%, 30 million are Americans who make up the top 12% of their own income distribution). On one level, this is a reminder that, as Thomas Piketty has observed, Europeans (and Americans) tend to underestimate just how wealthy they are. But on another level, it says rather a lot about the sort of economic policies that have been pursued over recent decades. Be it outsourcing jobs or the drive to lower corporation tax, these policies have entrenched the position of the wealthiest in their citadels while disaggregating the rights of the poor everywhere else.[18]

A less dramatic but more pervasive problem with inequality when it reaches this sort of level is the negative impact it has on social mobility:

the bread and butter of the American dream. At the end of the Cold War, about one in four of the world's inhabitants lived in areas with improving standards of living, but today that same figure has fallen to one in six.[19] At the same time, seven out of ten people in the world live in regions of growing inequality.[20] In modern society we almost all expect to do better than our parents did. So when sons and daughters grow up to be the same as, or worse off than, their parents, resentment lurks in the future. Again, this is a global phenomenon (for all that it may be somewhat unevenly experienced). Wealthier citizens will encounter it as a struggle to keep up with the Joneses; its effects will be psychological—managing disappointment will become a valuable skill. For the world's poorest, it will be more about keeping their head above water.

When people start to feel that they are doing less well than earlier generations, and above all if they begin to feel there is simply no point to playing by the rules, then the promise of social progress and just rewards that underpins modern society begins to fray at the edges. At present 82% of parents in China think their children will be better off than them; in Brazil, 79% do. Yet the longer-term trends are likely to disappoint these parents and their children even more so, because both countries are "greying" at least as quickly as they are growing: by 2050 a full quarter of the population of Asia will be older than sixty. And as citizens of the West are only too aware, someone needs to pay for those who have retired.

Inequality is one of the few things that it can be fairly said falls to all of us to try to fix. Inequalities in rich and poor countries may manifest rather differently: as a vertical problem of class, say, or a horizontal problem of ethnicity. But they everywhere are inefficient. They everywhere entrench the misery and suffering of poverty at the bottom end of the social scale. And they everywhere encourage a sense of rightful privilege among the wealthy that they somehow live apart from the rest: subject to different rules, exempt from the mundane tasks it falls to others to perform.

With less and less reason to believe that they have very much in common with others at all it becomes all too easy for the rich to believe that to them alone fall the higher duties of management and global leadership, the noblesse oblige of the red-eye flight, corporate luncheons, and signing off on the annual reports. It becomes all too easy for the rest of us to acquiesce in a system where we must struggle against one another to get their attention, and where the poor of rich countries are made to feel like the soup kitchen is a privilege compared to what others must content themselves with (though, of course, they can always aspire to be promoted onto "benefits").

Inequality is indeed incendiary. It is too obviously a problem for democracy (let alone human dignity) that, as in the figure here, the income of the top 1% globally is about the same as the world's poorest 4 billion people combined; or that the richest 0.14% of the world's population owns 81.3% of the world's net financial assets. It matters no less that more than 2 billion people struggle to survive on less than $2 a day, when they could be doing so much more with their time. And we should all feel a nip in the gut when we hear that one person in eight goes to bed hungry every night.[21]

And yet the world is not, of course, made up of just the Fortune 500 on one side and the global poor on the other. What is happening (or rather, not happening) in the middle of this picture is also important. Globally, 1.8 billion people earn between $10 and $100 a day, to take a not-uncommon measure of global middle-class status. By 2020, this will reach more than 3 billion people, more than half of them in the Asia-Pacific region alone.[22]

This emerging global middle class is often looked to, contradictorily, either as proof that global capitalism really does see us all good in the end or as the hoped-for fountainhead of resistance to that selfsame system (the poor are presumed not to have a voice of their own about such matters). In reality, though, this "global middle class" is simply a class of people who will, for the foreseeable future at least, be spending their marginally increased income on a slightly better set of the basics of life. Their new spending habits will not stop inequality from growing at the extremes.[23] And their marginally improved status may not even be enough to guarantee that they themselves remain clear of the rake of the poverty lines in their own countries.[24] The situation looks most promising in Asia-Pacific. But in Africa, many of those who are counted

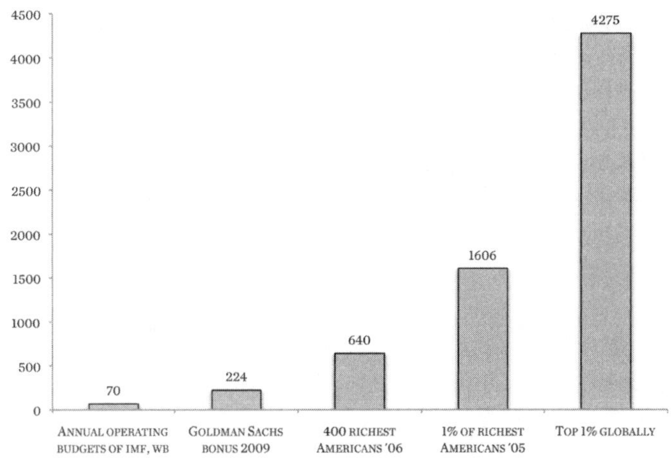

FIGURE 5. "Some incendiary statistics: Incomes of the rich expressed in income of the millions of poorest" (after Branko Milanovic).

as "middle class" continue to work in the informal sector. Their existence as a class is tentative, to say the least: they are still each living off about the same as a European cow.[25]

Among the more politically fraught facts of inequality are some of the more direct ways in which the consequences of growing global divides are affecting the rich world too. Because it isn't just the developing world's middle classes who have been losing out in recent decades: the rich world's middle classes have gained less than anyone from globalisation. Their plight, Milanovic shows, is matched only by the absolute poorest in the world. The rich world is also affected by the rise in migration from places offering little or no prospect of a livelihood to places that do. It has become routine within political debate in Europe and America to ignore the actual history that drives such migration (after all, overlooking the reasons people migrate is essential to sustaining the chauvinism of most anti-immigrant policies). But the *geography* of international migration is very clearly being taken into account.

This is apparent at every border wall, many of them militarised, that exists or is currently under construction between lands of poverty and plenty: be it the land border that limns the Spanish enclaves of Ceuta and Melilla in Morocco (the seventh most income-unequal border in the world), the electric fence Botswana is constructing along its border with Zimbabwe (the tenth most unequal border), or the US-Mexico border (the seventeenth most unequal border), scene of what President Obama recently termed a "humanitarian crisis".[26] This gradual fencing-in of the rich world is what politics has learned from laissez-faire, neoliberal economics. It is what my colleague at Queen Mary Gerry Hanlon means when he talks of "management as symptom", of the tail wagging the dog as the rich seek to confine the problems of inequality to other people and other places, albeit at the cost of making life that much harder for everyone, themselves included, in the process.[27]

This emergent geography of barriers and fences is more than just a tacit acknowledgment that the world overall is more unequal than any single country is. It is one of the reasons the world is so unequal. And it is no cheap fix either: citizenship tests, fast-track business lanes, weaponised border walls, and the growing twenty-first-century archipelago of asylum and detention centres connected by privately chartered jets are all horrifically expensive. And none of them will secure the "democratic character" of any nation, anywhere, for all that someone like Benjamin Netanyahu would have us think so.[28]

We are fools if we continue along this path simply because we really have come to believe that there is no alternative. Some would argue, of course, that a certain level of inequality is desirable: why should we work hard, or work at all, if we are not properly incentivised to get ahead? And it is true that a degree of inequality is inevitable. But that does not make it desirable, or its growth something to celebrate. Neoclassical economists may live in a world where competition is, in the last analysis, all that really matters. But most of us fortunately do not.

None of us is free from the struggle over geography, however, as Edward Said once put it. And as he had the wit to also see, that struggle is as much about ideas and imaginings as it is about soldiers and cannon.[29] Here we may all fairly be said to be finding ourselves on the losing side today. In Britain, the old Poor Laws that preceded the Victorian workhouse recognised that, because the poor often had no choice but

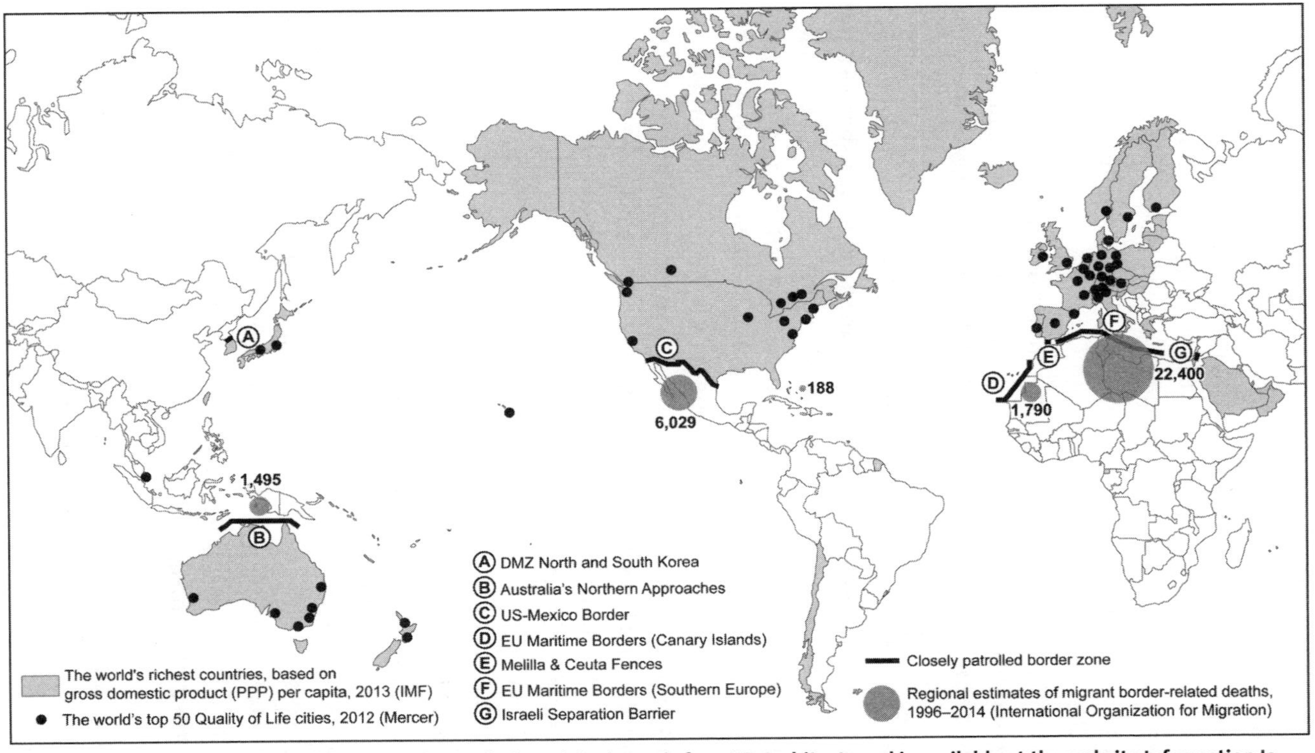

FIGURE 6. **Keeping out the poor (after TD Architects). The original map is from TD Architects and is available at the website Information Is Beautiful (http://www.informationisbeautiful.net).**

to move to find work and survive, and because on the whole they had more than enough reason to want to work hard, what they really needed was not additional motivation but portable welfare, to keep them in good health when work was harder to find. It is a measure of just how accustomed we have become to high levels of inequality to consider that a policy supported by the English Whigs in the eighteenth century would be deemed "radical" today.

But it is not just the growing gaps in income that require that we tackle inequality head on. Most of the so-called emergent diseases of globalisation—bird flu, swine flu, HIV/AIDS, and the growing scourge of multi-drug-resistant tuberculosis—are also actually diseases of inequality. It has been known since pioneering research in the 1980s that HIV infection rates are greatest wherever inequalities are sharpest (just look to post-apartheid South Africa, where prevalence rates top 40% and the Gini index for inequality sits at over 0.65). The appearance of swine flu in Mexico (it was predictably referred to as "Mexican flu" for a while) turned out to be a product not of Mexican animal husbandry standards but of American-owned confined animal-feeding operations, which had relocated to Mexico to take advantage of cheaper wages. So for all that Mexico was itself pathologised in the first instance, a better diagnosis might after all have been simply to label it "NAFTA flu".[30]

Indeed, it seems likely that the biological risk factors leading to the species hopping of influenzas more generally are a consequence of the economic pressures that poor farmers in countries like Mexico and Vietnam are put under by wealthier nations, including middle-income countries like Brazil, in their demand for cheap protein in their diet.[31] At the same time, the global production and supply of vaccines such as Tamiflu is heavily skewed towards the rich world's consumers (to the tune of their possessing 95% of available vaccine supply), even though these are the countries that stand some way back from the front lines of viral exposure.[32]

The geography of such problems as MDR tuberculosis, which particularly affects unequal nations like South Africa and Brazil, is a further reminder that the burden of inequality in global health falls squarely upon the poor world (where it is harder to maintain the drug supply, and where other associated diseases, such as HIV/AIDS, are more prevalent). But the rich world should not consider itself entirely immune. In 2013 one of the hottest reported zones of MDR tuberculosis was precisely the US-Mexico border: a reminder that, as one American researcher put it, "we're all connected by the air we breathe."[33]

Yet we are connected too by the political systems we live in and the rules they lay down for us about what we can and cannot do in relation to others. We have grown accustomed of late to celebrating individual freedom—what the British philosopher Isaiah Berlin once called "negative" liberty—as if it were an immutable right. Since the beginning of the fall of communism in 1989, we have not seen fit to consider even whether there may be a trade-off between this individual freedom and other social goods that we require to live safely and well. Responsibility was the way of the oppressive state, we are told, and that inglorious project hit the sands when the Soviet Union fell. But in our refusal to countenance any limits at all on our individual freedoms, a quarter century later we have simply traded in the nanny state in exchange for Big Brother. Why? Because inequality, which is the product of untrammelled freedom *without* responsibility, takes away from our societies more than it gives back to any one of us as an individual.

THE PAST AND OTHER COUNTRIES

When confronted by global inequality in the abstract, one is forgiven for thinking that, like globalisation, it is something beyond our control: too diffuse in its causes, too widespread in its effects. But examined in a little more detail, and the political choices that stand behind it soon begin to take shape. Examined in any one country over time, and it is clear that there are things beyond the inevitable iron laws of capitalism that influence the nature and extent of inequality. It is these things we can take hope and a little direction from as we confront the sheer scale of the problem internationally.

Given that there is now ample evidence that inequality will catch us all out in the end, it is high time we did this. Inequality, we have seen, works counter to the prosperity our politicians pledge to deliver. It works counter to economic growth, as even the IMF has come to realise belatedly.[34] But above all it works counter to democracy. The people whose interests it more consistently serves are the strong arms and the oligarchs, be it General Pinochet in Chile, Hugo Chávez in Venezuela, or the plutocrats in present-day Russia.

As R. H. Tawney wrote, "It is the nature of privilege and tyranny to be unconscious of themselves, and to protest, when challenged, that their horns and hooves are not dangerous, as in the past, but useful and handsome decorations which no self-respecting society would dream of dispensing with." Privilege and tyranny are common enemies indeed today. They create, as Tawney wrote, a culture of domination and callousness on one side and resentment on the other. Above all, they foster a sense of "suspicion and contention" in society at large.[35] Little wonder, then, that "trust cannot thrive in an unequal world", as the political scientist Eric Uslaner has argued.[36] But the opposite of this, that inequality thrives in an untrusting world, where the values of mutual respect are lacking, is equally true. The question is: what can we do about this?

One obvious starting point is to try to understand the reasons why at present we tend not to trust one another. As it happens, when it comes to questions of poverty, these reasons have proven remarkably durable down the ages. In nineteenth-century Britain, the underclass was often described as coming from "another country". Those in today's global underclass, who *do* come from other countries, are described as if they were still living in the past. This is the ultimate condemnation, it would seem, since our instinct is to assume that this renders them separable from us: endowed with other fates and less sophisticated desires, hardened to bear what nature brings.

Our own poor, meanwhile, are denigrated as "chavs" (in Britain) or as "RMIstes" (in France). In the United States the overtly racial bracketing of the poor does the denigrating for us.[37] The distancing effect is in every case the same. Yet there were different ways of thinking about poverty in the past, many of them articulated by the poor themselves. "Our lives shall not be sweated from birth until life closes; / Hearts starve as well as bodies; give us bread, but give us roses", as America's Lawrence mill-girl strikers put it in 1912.[38]

The very first systems of nationwide social insurance that thinkers like the Marquis de Condorcet argued for in the early to mid-nineteenth century, and that the Lawrence mill workers were striking for half a century later, were not driven by notions of charity. In many cases it was actually a conservative fear, which Marx shared in the form of a radical hope, of the dangers of the molten mass being left to get just a little too hot beneath their oily collars.[39] But that fear was also a recognition that the poor really did form a part of the same society and that addressing poverty meant self-critique by the rich as much as "becoming deference" by the poor.[40]

If the world today is a different place, it is in part because critics and reformers in the past challenged this status quo with ideas that seemed radical and unrealistic at first but eventually began to settle in the public imagination as social truths requiring action. "It ought not to be left to the choice of detached individuals as to whether they will do justice or not," wrote Thomas Paine to a passive American public in the late eighteenth century—just after the American Revolution.[41] And though it took most of the following hundred years to realise that vision, it was ultimately because popular sentiment *did* change that, by the end of the nineteenth century, over any given three-year period, as much as a quarter of the American population was receiving some kind of poor support. That same change is why the majority of Americans today also receive some form of welfare support at some stage in their lives.[42]

Without these systems of social assistance, and at least some institutionalised commitment to the value of equality that sustains them, Britain and America and the rest of the rich world would be very different places today, however successfully those nations also innovated and exploited their way to great wealth (and however keen they now seem to be to throw those gains away). Yet we have today become accustomed once more to the moral and material convenience that intellectual passivity brings: content with the occasional well-meaning flap about "the life we can save", content for the most part with the way things are. Popular sentiment, it seems, has yet to take the step globally that today's advanced countries took nationally a little under two centuries ago.

World poverty as it confronts us today is no less part of a wider social question than was poverty in rich nations in the past. And global wealth is every bit as central to the constitution of that problem now as it was back then. Throughout history, societies began to address the problem of poverty at the very moment they also started to recognise that it was a problem involving the rich and not just the poor. Of course, the obstacles to acknowledging this globally are now that much greater. Not least, the language we require is one we have been better at speaking before, more accustomed as we once were to its vocabulary of equality and fairness. In the past such words always had a distinctly national flavour. We will need new and even bolder thinking today if we are to reinvigorate these words and put them to work internationally as deeds.

There is cause for hope, however. It is perhaps too easy to forget that less than a hundred years ago progressive rates of national taxation and pension schemes were all largely unheard of still in the rich world. It is even easier to forget that scarcity—of something so simple as even paper—and the rationing of basic food items was common in those same countries just seventy years ago. And in forgetting this, it is all too tempting to assume that the realisation by a broader swathe of humanity of the social gains we have since made (along with the political rights upholding them) is unrealistic, utopian even.

Their doing so is not unrealistic. It is imperative. But "development" cannot be the way of it. Aid and development have become increasingly professionalised and technocratic: the surest possible sign that they serve vested interests more than those of the poor themselves. Today's well-meaning development industry equally obscures the historical responsibility for colonialism that should properly be borne by the very nations that are today offering overseas assistance. After all, there are just four countries that are considered poor today that were not also once colonised by today's rich countries. The achievements of development workers over the years ought not to disguise this.

This merely reminds us once again, however, that what is needed to address global inequality is a *political* disposition first and foremost. As Michael Ignatieff reminds us, the wellspring of all politics is "located in [the] human capacity to feel needs for others".[43] It may be true that the desire to meet and to determine human needs is a utopia too readily conjured, the slide from a discourse on social graces to the gulag too well documented. But we cannot continue to hold hostage to just one of the twentieth century's numerous tragedies our growing twenty-first-century need to feed ourselves properly and to clothe ourselves with dignity and respect. If the world's rich were prepared to learn from the poor rather than always lecture them, they would see this. India has a long tradition of anti-corruption and of thinking about the good life; Latin America has a lot to teach about the role of public virtue in political debate. There is scope for a great deal more public and political innovation in our world.

Questions of public virtue—of distribution, equality, social policy—are greeted as passé in the West today, especially when it comes to international affairs. Such questions are disliked by mega-donors such as Bill Gates who would rather keep politics out of things altogether. They are also alien to the technocrats at the IMF and a good many NGOs who are trained to deal in focused and measurable solutions and not to think about the bigger picture: whether the penny they are throwing is even going into the right pond. These questions are feared perhaps most of all by governments for the simple reason that they demand the more politically risky strategy of challenging the privileges of the rich as much as the habits of the poor. This is short-sighted. We know from history that inequality is not necessary and, with political intervention, can be addressed. We gain at least a sense of this from the two figures here, which chart pay dispersion within manufacturing for selected countries over time. Both reveal how the gap in pay between rich and poor rises and falls in close step with substantial shifts in policy and the wider political context.

And so we are back where we started: the persistence of great poverty and inequality around the world has less to do with what the poor world presently lacks than with what the rich world once took from it or prevents it still from getting. To think otherwise, to assume that poverty can be alleviated by the uncertain benefactions of philanthropy or the latest intellectual fad to hit the NGO sector, is to fall prey to what Gunnar Myrdal once aptly termed "mental conspiracies".[44] Global poverty has less to do with market failures than with the successful exploitation of

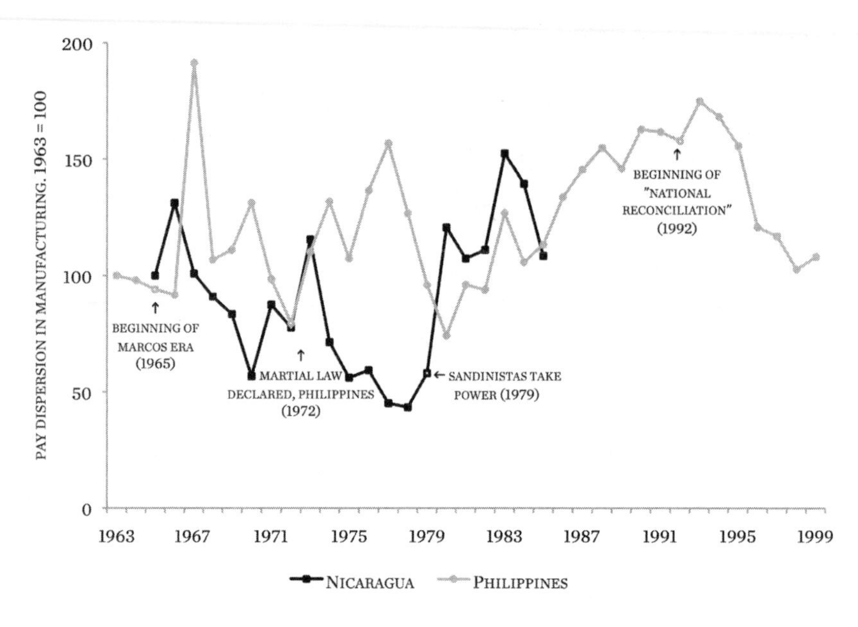

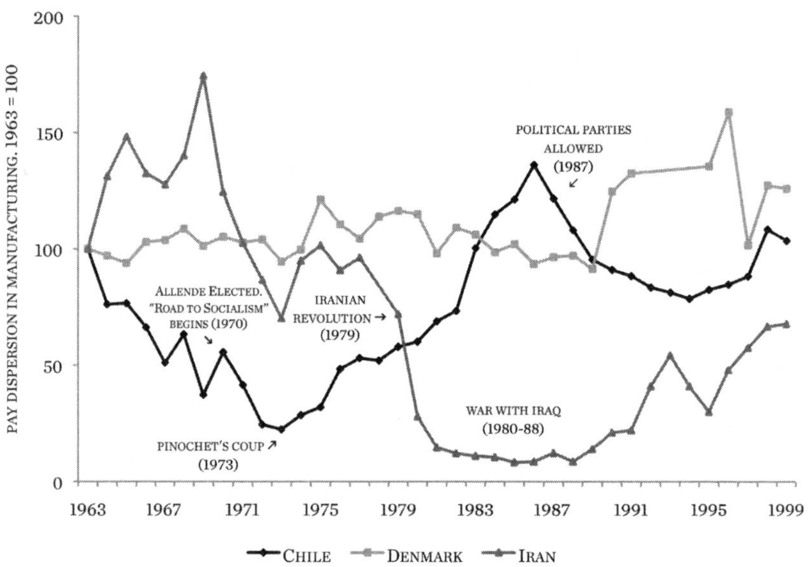

FIGURE 7. Macroeconomic causes of inequality change (after J. K. Galbraith).

markets by the powerful. But above all, it has to do with the political atrophy that results when those of us who might actually hold those powerful to account accept this—and accept, in doing so, a world that achieves but a fraction of its potential—in exchange for a quieter life.

The quieter life is not always and everywhere the better life, however. Until we recognise this, until we recognise our own complicity within such structures as those which sustain the current state of affairs, then we will not "solve poverty in our lifetime", as Jeffrey Sachs assures us we can. Rather—as T. H. Marshall more aptly put it—we will simply keep "abat[ing] the nuisance of poverty without disturbing the pattern of inequality of which poverty [is] the most obvious unpleasant consequence".

THE CASE FOR EQUALITY, GLOBALLY

Actually addressing the extent of global injustice that inequality bequeaths will involve a much broader, politically minded discussion than we have been having to date about development and globalisation, international aid and trade regimes, emerging markets and their promise of new middle-class consumers, or indeed any of the other component parts upon whose collective interaction the persistence of global poverty depends. Above all, the discussion must encompass the real history behind the entrenchment of a partial and frequently unfair international order and address head-on the structural injustices that underpin that order in the present.

As it happens, the general public is not as naïve to this history or this sort of reasoning, as social scientists and political commentators sometimes think. A 2005 poll reported that more than 80% of Americans *already* thought inequality was a major problem in the United States, and that was before the financial crisis hit headlines. Just as interesting, 60% of those surveyed also thought it was the government's task to reduce inequality. We can conclude from this only that there must be a good many disappointed Americans today and that Americans' frustration with the government is not always for the reasons the Tea Party would have us believe. But we might also conclude that the language of equality is one that we get better at with practice. So should it also be internationally.

We know, for example, that more equal societies give more back, both to their own communities and to distant strangers.[45] So it is from figures like the one here, showing the relationship between foreign aid and income inequality in rich countries, that a more hopeful hand is extended. But something other than aid is required as well. We need to commit to addressing the underlying origins of inequality, not simply to reducing its effects. This is not a task that falls to rich nations alone. More equal countries are better able to convert the GDP growth they do achieve into actual poverty reduction, so reducing inequality in poor countries is equally important here.[46] But above all we need to find ways of breaking into and loosening up the inequalities of the mind that continue to hold rich and poor the world over in an intimate and yet ultimately rather unloving embrace.

Many prefer not to see things this way. "The juxtaposition of western affluence with third world poverty," says the usually sage Kwame Appiah, "can sometimes lead activists to see the two as causally linked in some straightforward way, as if they are poor because we are rich."[47] Appiah is right to be sceptical of condemnations of capitalism (or even self-interest) tout court, but his diagnosis is quite wrong. The growing wealth of the richest in society does often come at the expense of more vulnerable others.

Competitive deregulation and wage reduction in poor countries today are driven by rich countries' policies and demand in the first instance, just as during the nineteenth century Irish labourers came to England and undercut the wages of the English working poor, because they had no choice but to flee famine in their own country (a famine the British themselves knowingly brought about). It was W. E. B. Du Bois who pointed out more than half a century ago that the marginalisation of the poor in the rich world and the poor elsewhere cannot be unrelated. He was right then, and he would be even more right today.

But to understand why, we need to look afresh at the usual reasons given for the poor remaining poor and what these reasons conceal about the ways the rich keep getting richer. We need to better understand the relationship between wealth and want in both its national

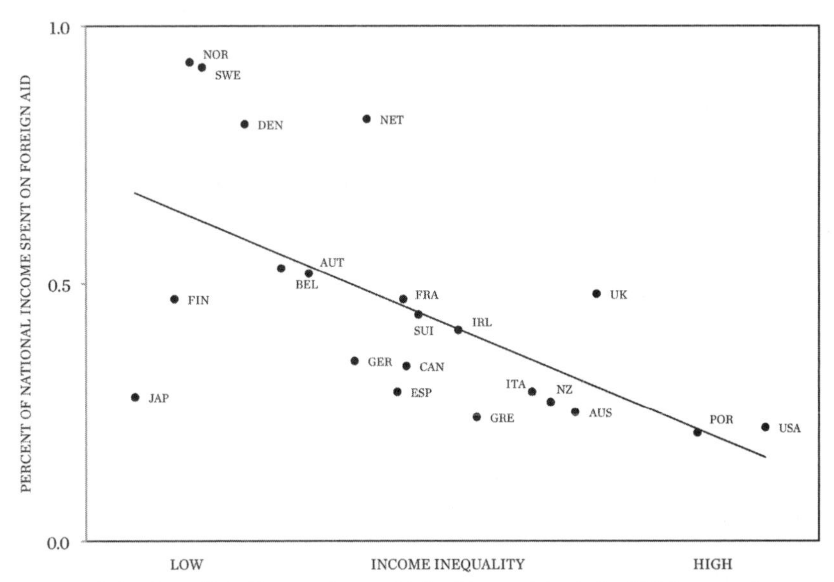

FIGURE 8. Foreign aid and income inequality in rich countries (after Wilkinson and Pickett).

and its international dimensions, because it is here that the political origins of inequality are ultimately to be found. For the same reason it is also here that the beginnings of a more equal world may be discovered. Where is our global New Deal? Where our *Rerum novarum*? To date there has been almost no political debate about this at all—although the one leader to have come closest is in fact the present pope, Francis. It is time to now have that debate.

As it stands, we seem to have lost all belief in the possibility of mutual aid, as if our needs were less mutually entwined than our desires. We have embraced a politics of security and risk in place of a culture of reciprocity and trust. And at the very moment when the problems we confront as humankind so clearly exceed the competence of any one nation, international talking shop, or individual to address, it seems we no longer see the value of thinking in the round.

Content to read just half the map, we have gotten lost and wandered off into the margins. We are unlikely to find our way back while we cling still to the prejudices of the past: the belief that some lives are worth more than others; the belief that nothing should remain beyond the market; the belief that the right to life, liberty, and the pursuit of happiness is a privilege accorded just to some; the belief that it is only the most privileged of all who should have a say in any of this.

We need to rekindle the idea that a fairer world is possible.[48]

2

A GREAT DEBATE?

What thoughtful rich people call the problem of poverty, thoughtful poor people call with equal justice a problem of riches.
— R. H. Tawney, ***Poverty as an Industrial Problem***

What is it that stops us from thinking that a fairer world is possible? The American writer and intellectual Susan Sontag once wrote of the need to recognise how "our privileges fall on the same map as their suffering."[1] Yet the map of the world, we have seen, does not gather itself up as straightforwardly as Sontag suggests. Wars are today fought less often between nations than within them. Most of the world's poor now live in middle-income countries. More and more people are growing old in places far from where they were born.

Still, Sontag's basic point holds. Something is seriously wrong with the way we have come to think about the relationship between wealth and want, about which privileges are defensible in the face of others' suffering. Consider the award-winning advert for the luxury brand Louis Vuitton. Shot by acclaimed photographer Annie Leibovitz, the photo features rock star and aid activist Bono along with his wife, the designer Ali Hewson.[2] It forms a part of Vuitton's Core Values campaign. And though it does not itself focus on global suffering—it is an advert for luxury bags after all—everything being co-promoted in the advert, from local charities to fair-trade companies, does. Their presence (and co-promotion) is intended to endow the Vuitton brand with their moral capital—to define, in that way, what Vuitton's global values are.

Reflecting the exquisitely tailored bags the company seeks to sell, the photograph is perfectly crafted in pitch and palette as it updates the Victorian epic

FIGURE 9. "Every Journey Began in Africa": Bono and Louis Vuitton.

of discovery for the present. "Every journey began in Africa", the words tell us (somewhat superfluously, since the mise-en-scène leaves little room for doubt). As to where in Africa, that seems not to matter: the contrast is the thing. To such ends Leibovitz has arranged a basic wash of taupe vastness, conveyed by limitless grass and sporadic bush, over which the eyes of our explorers Bono and Ali roam free. The scale of the scene and the continent it represents is further amplified by the presence of the white- and red-enamelled plane. Before this rich-world stage effect, the two passengers grasp Vuitton Keepall 45s, co-designed by Ali herself. In the background, the sky roils up in a painterly cloudscape.

The question of poverty in all of this is implicit, metaphorical, and heavily sanitised: there are no other people, for one. But Africa and its struggles are essential to the picture: as the presence of an African backdrop, in conjunction with the protestations of core values, attests. As ever in the colonial imagination, although the natives may be important, it isn't appropriate for them to be cluttering up the background. Instead, what we have are a few feathered Kenyan charms hanging from Ali's bag. In compensation for the lack of a native presence, they offer a visual punctum of African lore, one that draws us to the main point of the photograph. Because these charms, it turns out, are made by fair-trade company MADE and are the defining feature of the Vuitton bags that Ali has designed (in fact, they may be the only thing that makes her bags different from any other Vuitton bags).[3]

Once we have digested all this, our eye drifts down to the accompanying text. Here we read that profits from the sale of both Louis Vuitton and the (third) brand being co-marketed here, Ali's own ethical clothing brand Edun (49% owned by Vuitton, it turns out), will benefit the local charity that she and Bono run in Uganda. Their appearance fees will benefit Chernobyl Children International.[4] Has ever the world been so generous? For all that it appears to be selling a leather bag, through its narrative turns this advert is in fact selling you, reader of the glossy magazine, the belief that you can make the world a fairer place, and then so beautifully. "Buy luxury goods, and you too can help save our planet," it says: a message that seems to be increasingly common. Witness Sean Penn and Giorgio Armani doing—well, doing what, exactly?—on behalf of Haiti and Penn's own charity, the Haitian Relief Organisation.

To put all this more precisely, these images of Bono and Penn would have us believe that the problems of poor people elsewhere in the world not only are unrelated to our own pursuit of wealth; they might even stand to be alleviated by it. It is imperative that we understand how we have come to think like this. To be sure, Bono in particular was a central figure in mobilising engagement around the anti-apartheid struggle in the 1980s and debt reduction in the 1990s. His conscience is no doubt cleaner than many of us can claim. But the inconsistencies he embodies are important. And he has worked hard to become a global symbol.

The world is becoming more attuned to such inconsistencies. This was made apparent to Bono when, in his other guise as a besunglassed singer, he performed at the Glastonbury Festival in 2011. There, a group of activists from the British group Tax Justice Network had planned to peacefully protest Bono's having moved offshore the accounts of his band U2—this at a time when the Irish economy was facing a possible default and its own poverty-inducing austerity. As the Edge started up on his opening guitar riff, the activists pounced, erecting an inflatable column with the words "U Pay Tax 2?" emblazoned across it, only to be immediately muscled away by security personnel.[5]

FIGURE 10. Sean Penn and Giorgio Armani, for the Haitian Relief Organisation.

Thus the point that Tax Justice Network sought to emphasise: drawing attention to rich folks' desire to secure the upper reaches of their own wealth, even when this is likely to be detrimental to society at large, is not only unacceptable dinner conversation; it doesn't even wash at a rock festival. Such things matter rather directly to the poor of this world, however. Taxes, as we shall see, are a central part of any functioning state and the wider political and moral economy into which that state is set. They not only provide funds but also impose discipline on governments and create responsibilities for providing public services. The general disparagement of taxes over recent decades and the failure to raise the tax base across the poor world are not incidental to the reason we have yet to rid ourselves of the blight of global poverty: they are in fact quite central to it.

We are all, of course, entitled to a good deal of self-interest, and what other way than by using their own celebrity status are well-known actors or musicians to respond to something they feel moved to act upon? But neither self-interest nor the peccadilloes of celebrity-cum-political mannequins are the heart of the problem. It is a more general loss of perspective that ails us. At the 2005 G8 summit in Gleneagles, Scotland, then British chancellor Gordon Brown hailed as a "historic breakthrough" the $40 billion of debt cancellation that had been secured internationally, through many gritted teeth.[6] Never before had anything like so much money been made available to poorer countries, and Brown was right to celebrate an initiative that had taken years to get to the table. Some of the personalities just mentioned helped mobilise the pressure to get it there.

Yet just a few years later, as the credit crisis took hold, the world's largest economies—with Brown leading the way, now as prime minister—moved swiftly to deploy more than US$1 trillion to save some of the largest Western banks, many of which had acted far more irresponsibly with other people's money than the poor nations whose creditworthiness and fiscal responsibility had been so fussily picked over at Gleneagles. And it is revealing that it was the bank bailout, not the debt write-off, which led Brown to fire off his unfortunate "we saved the world" comment.

POSITIONING THE POOR

If we are to do better than this, and we must, then we need to think far more *critically* about the relationship between wealth and want. As Martin Luther King Jr. famously put it: "In this unfolding conundrum of life and history there is such a thing as being too late."[7] But there is such a thing as the right time too. And we have reached, it seems to me, a much more significant turning point than that presented by the need to devise a post-2015 agenda for global development. Quite what that turning point consists in is the subject of the final chapters of this book.

First, though, we need to understand why at present we do not see this turning point for what it is. And to do that we must bring the problem of wealth more fully into the picture. We need a political analysis of the problems of a "wealthist" society—an analysis of the way that social norms and institutions are geared towards the uncritical promotion of wealth "at all costs" (not an analysis, much less a witch-hunt, of the rich, per se). And we need this so as to compensate for our somewhat overburdened analyses of the problem of poverty.

Consider the following question: What is at stake in the problem of world poverty today? It is not, as we are usually told, how the actions of poor countries may best be optimised so as to speed their transformation into something resembling our own states—especially since those "achievements" are now more in doubt than at any time in the past half century. Instead, it involves a wider set of questions we must all ask ourselves about the kind of world we wish to live in. Which sort of rights, such as freedom of movement, are we prepared to ensure for all? And what are we to do with those of our own current privileges that may clash with those objectives? Bite-sized solutions will not cut it when it comes to global hunger. As William Beveridge observed when introducing postwar Britain's welfare state, "A revolutionary moment in the world's history is a time for revolutions, not for patching."[8]

Yet "revolution" is a serious word: one best kept in a glass case with the keys in another room. Take the Cuban Revolution: an obvious case for anyone suggesting that poverty eradication be achieved by tackling wealth rather than want. The Cuban Revolution is Susan Sontag's thesis, but with guns, and indeed butter, if ration cards are your thing. But the fact remains that for all that the revolution has reduced wealth and income inequality, it has done so coercively (wealth snakes might not be better than development ladders), and the Cuban government cannot blame all of the economic stagnation under which its people labour on the US embargo. But if not actual revolution, about which we take our cue from comedians these days anyway, then what?*

*Russell Brand, *Revolution* (London: Century Books, 2014). In fact, it is Thomas the butler from *Downton Abbey* who had it right: "Not a revolution but justice for the majority," as he put it in the first episode of season 5.

The laissez-faire pragmatist, to take the other end of the political spectrum, will tell you that there is nothing more to be said or done here because "always was it thus"; the poor have always been with us. Let us leave to one side the fact that it is just this sort of cynicism that revolutions thrive on. The point, rather, is that the situation is neither as irreparable as revolutionaries profess ("Off with the king's head!") nor as perfectible as those who benefit from existing arrangements like to claim ("Patience, dears, patience"). The question we should really be asking ourselves, then, is this: just what, exactly, do wealth and poverty mean to us, and what, if anything, we are prepared to do about this?

We might start with poverty first of all. Poverty takes different forms depending on what a given society chooses to recognise as such and whom it assigns to that category. This is what the historian Lewis Coser was getting at when he wrote that, while we indeed have always had the poor, "the poor have not always been with us."[9] In the Victorian age, the "poor" who had been centre-stage in a long, drawn-out debate over reform of the Poor Laws in Britain in the 1840s were then largely forgotten, politically speaking, until their rediscovery, on quite different terms, in the 1880s—this time as moralised subjects like Dickens's Work'us. It took until after the Second World War for a new type of poor to be "discovered", out there in the newly independent nations of the world. This was when the poor of other countries could no longer be ignored by an emergent *international* agenda. It was, as the anthropologist Akhil Gupta reminds us, when the very idea of today's global poor was planted, the seemingly tectonic fastness of their plight established.[10]

This new category of the poor was always too large for anyone seriously to get a handle on, however. As the moral philosopher Thomas Pogge points out, there were millions killed in the Irish potato famine—An Gorta Mór—as well as under Stalinist forced collectivisation and during Mao's Great Leap Forward, and blame was easily attributable in each of those cases. But the problem of *world* hunger—and global poverty for that matter—is that for all it concerns a comparable number of victims alongside these other atrocities, its causes are simply too diffuse, we are told.[11] Yet in this very diffusion also lay the temptation to craft this most recent category of the poor into whatever it is we wanted to see whenever it was we happened to look.

Accordingly, the wealthy world's images of the rest of the world's poor were made to fit the political needs of the time. For much of the post–Second World War era, that meant that the poor were trussed into the straitjacket of Cold War calculation. But not before they inherited the hand-me-down prejudices we were beginning to see didn't always fit our own poor. The same took place through the prism of the market in the post–Cold War era, when the global poor were of interest as a potentially expanding mass of consumers jostling together in "emerging" economies. "If we can't fix their poverty, if we can't bring justice to them, perhaps we can at least bring them to the market", ran the logic. Accordingly, state aid declined, personal and private donations went up, the private sector strode to the front of the stage, and the poor at the edges were asked to accord to the vision the rich world now had of them as micro-entrepreneurs (such fulsome praise!) in the making.

Today, when it seems that this does not work either, we lose patience with these poor whom we have created. We throw our hands up and say, "Aid does not work!" We treat them as "failed consumers"—which, given the priority we accord to consumption, is practically to deny them their existence.[12] Within this new moral economy of the poor, writers like Robert Kaplan take on the role of modern-day, global flâneurs. On our behalf these writers roam the darker, hidden spaces of the globe, taking us into the slums of Rio and Mumbai and the lawless reaches of Pakistan's Federally Administered Tribal Areas, crafting a global imaginary of famine and humanitarian crisis as they go. But in so doing, they also collapse notions like poverty and threat together, encouraging the rest of us to look more favourably upon our governments as they send our troops to police those places: all of which gives our correspondents yet more reasons to book the next flight out there.

The term "global poverty" is today tacked on to and gives shape to this history. And like its first cousin, "global health", it does not do so apolitically. To draw from Akhil Gupta again, the frequency of use of the term "global poverty" increased five-fold from 1999 to 2005. Global poverty has thus become the altar of a very broad church. And yet we can find it quite easily on the map still: for its headquarters are at the World Bank and the International Monetary Fund in Washington, and it is codified in the Millennium Development Goals and mission statements of a thousand non-governmental organisations in Geneva and New York. What the wealthy world understands as global poverty today is largely the product of these organisations' understanding, which is to say the understanding of those (educated, Western) elites who mostly staff the upper echelons within them.

Unlike wealth, which always has powerful interests to protect it, poverty has no way of resisting the efforts (even when they are well intentioned) of those who want constantly to monitor and impose their reality upon it. This is as true abroad as it is at home. But it is not always visible, as with the narrowed policy freedom of poor countries that the IMF's Poverty Reduction Strategy Papers bring about.[13] It is highly pervasive, however. Take a tour of the regional and head offices of the main international organisations in Geneva—the UN Development Programme, the UN High Commissioner for Refugees, or any number of the NGOs and humanitarian organisations with headquarters there—and all are walking in step with the latest trend to be rolled out via their national offices around the globe. In Washington, DC, stop by the IMF or the offices of the President's Emergency Plan for AIDS Relief, or take a trip over to Seattle to the new Bill and Melinda Gates Foundation building, and the story is just the same.

These are all organisations doing the best they can to address problems that are not theirs for the solving and that they can address only by converting the impossible situations they confront on the ground into what look like "doable" programmes when sketched out on a whiteboard at headquarters: bring this many bed nets to this many people in this identified area, report back to line managers, advocate for the community, and seek funding for more. The cycle continues, as, most usually, do the problems themselves. There are countless thousands of lives improved along the way, and yes, even "saved", though we might quibble with the language: some diseases have very nearly been eradicated, and millions have been moved out of poverty. But these are the showpiece examples. Speak to people within these organisations off the record and few will deny that most of what they do is merely the best they can with the resources available, that in a fairer world they would be doing something else.

But a fairer world we do not have, and the manner in which the modern international organisation sets the international aid agenda is increasingly a manner learned and adapted from the corporate sector: just look at where many chief executives of today's major aid and humanitarian organisations did their time before taking up the call of what the comedian Paul Whitehouse once memorably mocked as "cherridy".

The result is precisely the overbearing "global values" talk today beloved by everyone from Louis Vuitton to the government of Norway: talk which closes off any solutions that do not conform to the latest buzzword fanning the interest of the international community (buzzwords from which that community derives its own power and influence). The right goes on blaming the poor, the left goes on blaming the right, and between them poverty carries on regardless: year after year, new initiative after big push, global compact after international challenge.

This is futile, of course. But there is a logic to it all the same—a logic to the way in which we constantly move poverty around, reconstituting it and affixing it to other people (the bottom billion, the landless, refugees) and other places (landlocked countries, aspirant nations, failed states). For just as in the past, by categorising and shaping poverty—by ensuring that it is defined as what other people lack in terms of the wealth that we enjoy—the privileged among us ensure that we are always the ones who get to set the terms of the debate.

A GREAT DEBATE

And just what is the debate we are having, exactly? Of late it has mostly been a discussion between learned men in suits (or cords) about the merits and demerits of aid. The front lines in this debate fall chiefly between those calling for big aid and those calling for no aid (if they are not declaring aid dead altogether). Those who argue for big aid believe that our generation has a choice today that our forebears did not. That choice is to give away enough of what we have so that others, who have nothing, can build up their own lives into something resembling our own. We can all make a difference, the moral philosopher Peter Singer says, by giving just a little bit more than we do.*

From each according to his means, then. But this says little about to whom, according to their needs, such money goes, and it says nothing at all about the structural injustices that determine why some of us have more to give to others in the first place. As Peter Singer would have it, Bill Gates should be praised for giving, not blamed for the way he uses wealth to meddle. In truth it might be better to leave praise and blame out of it altogether and focus instead on why the rest of us seem as happy to tolerate the growing ranks of the billionaires (their number increased three-fold from 2002 to 2012) as we are to accept that millions of children go hungry every day.[14] As Oxfam reported in 2012: "The top 100 billionaires added $240 billion to their wealth in 2012—enough to end world poverty four times over."[15] On one level it really is that simple.

This has long been the message of Jeffrey Sachs, perhaps the single most influential voice in support of big aid.† Unlike Singer (who puts his chips on individual morality), Sachs thinks it is technology that needs to be mobilised in support of big aid, and he thinks that the power of the state should be channelled to that end. For Sachs the state is best thought of as an effective delivery and enforcement mechanism: it is certainly nothing so grand as a democratic focal point for citizens. This is politically short-sighted: perhaps not surprising for one of the chief architects of the shock therapy programmes inflicted upon the people of Central and Eastern Europe after 1989. It is also the reason Sachs still feels able to talk, as he does in his book *Common Wealth*, of the "stabilization of the world population at 8 billion or below by 2050" as a concrete goal for humanity.[16] The ghost of social engineering past truly lurks behind statements such as these. Are we to issue permits? The problem here is that by reading poverty as a problem of the people who are the bearers of that problem, we can convince ourselves that we are within our rights to do all manner of things to them in order to address it.

*Peter Singer, *The Life You Can Save: Acting Now to End World Poverty* (New York: Picador, 2009).

†Jeffrey Sachs, *The End of Poverty: How We Can Make It Happen in Our Lifetime* (London: Penguin, 2005).

Sachs is, however, attentive to the fact that rich countries developed in ways that poor countries are now prevented—barred, in fact—from doing: fossil-fuel-driven industrialisation, protective tariffs and trade barriers, for example. But he still thinks of poverty as an attribute of the poor rather than the outcome of uneven and often unfair relationships between people: as a product of the slings and arrows of human misfortune, of how close to the equator one is born, of how dependent one is on natural resources.

Enter, stage right, the Millennium Village, Sachs's prototype for aid in Africa, which is to show (though to date it still hasn't) that if we get all the technical details right, the poor may count themselves on the road to salvation. The idea of Millennium Villages is both pastiche and unintended parody of that other cliché, the Global Village. It is perhaps no coincidence that the two concepts, unappreciative of politics as they are, have emerged historically more or less alongside each other.

Where Sachs is correct is in his recognition that free markets are never quite as free as their supporters claim. Sachs's greatest critic, William Easterly, does tend to think this, however. Easterly frames his influential critique of the big-aid agenda in terms of how well Sachs's claims stack up to the assumptions (though not always the realities, we should note) of neoclassical economics. And in Easterly's view, the answer is "badly".*

Yet Easterly himself underplays the value that careful aid can have (one thinks, indeed, of the Monty Python line "What have the Romans ever done for us?").[17] And just what is the future that Easterly offers us anyway? His support of the free market and his preference for leaving people to their own luck within it is the opposite of Sachs's top-down paternalism. Yet it makes no fewer heroic assumptions. Big aid is wasteful and the money never reaches the poor, Easterly says: $2.3 trillion has already been flushed down that hole. Experts are not the solution but the problem, he says, presumably in the spirit of the rank amateur.

*William Easterly, *The White Man's Burden: Why the West's Efforts to Aid the Rest Have Done So Much Ill and So Little Good* (Oxford: Oxford University Press, 2007).

States too are not the solution but the problem, and so until we do something about these various scourges, then the "rights" of the poor will continue to be abused.[18]

Yet the problem is not that poor people's rights go unrecognised. It is that they are recognised largely within structures that ensure they remain largely meaningless. The problem is not just the "unchecked power of the state" in poor countries but the not infrequently exploitative Western governments, unhelpful international rules, and unfair global markets that determine what those states can do. Easterly fails to see that the power of a developing-world state is invariably checked (if not checkmated) by the very market forces he would have us unleash more fully. And for all he points us to the lessons of the past to draw our morals, he gets his own history in a muddle: the history of the ideological and systematic abuse of the rights of civilians by the state is a European and American history as much as one pertaining to autocratic governments in the poor world; and half of the time, the market too ought properly to be standing as co-defendant to the charge.

But Easterly's are influential ideas, and the view that aid does not work, that "the debate is over", has become more popular of late. This has certainly been apparent in the response to another influential voice, more radical than Easterly and more uncompromising in what she thinks is to be done: that of onetime Goldman Sachs economist Dambisa Moyo.* Moyo concludes from the failure of Africa to develop that we would all be better off if we would just cut the aid lifeline altogether, the better to wean poor nations off a source of domestic economic distortion and disincentivisation.

This would be harsh medicine indeed but is anyway based on the faulty premise that the biases in international trade regulations (which do more to undermine poor countries' development than any failures of aid itself) are not a more serious fish to be fried first. A more nuanced version of the aid-is-bad argument comes from Princeton economics professor Angus Deaton. He rightly suggests that aid doesn't work be-

*Dambisa Moyo, *Dead Aid: Why Aid Is Not Working and How There Is Another Way for Africa* (London: Penguin, 2010).

cause, too often, it is intended *not* to work. It is used instead as a way to buy influence or access, or even just to buy off our own troubled conscience.*

Yet if aid is unavoidably political in the way that Deaton and Moyo allude to—and they are surely right about this—then all of these authors have yet to grasp the more fundamental implication of their own insights, namely that we start with the politics behind the making of global poverty and go from there, leaving aid as a side avenue to take up at an appropriate point along the way.

But this has not been the way the debate has been going. As the macro-debate on international aid has run into the ground, the approach has been to flip the coin and try all the same ideas again, this time from the ground up. The current vogue thus sees the lack of development as a problem best tackled at the local scale: be it microfinance schemes as are offered by Bolivia's BancoSol or, more famously, and not without its successes, Muhammad Yunus's Grameen Bank in Bangladesh.

The idea of local ownership is a fine one. But any idea, when it becomes a mantra, is a problem in waiting. This is most apparent with the turn to "evidence-based policy-making" within aid and development organisations. In shifting our attention to the question of local "impact", the evidenced-based approach to poverty reduction presents itself as a veritable people's flank in the age-old development wars: the poor sign on to more bite-sized programs, and in return they get to reclaim a bit of space from the behemoth NGOs and lumbering states that usually monopolise the field. For those running such schemes, this has the distinct benefit of making programmes and policies appear both more directly "pro-poor" and more democratic. And yet we have been here before, of course, with the justification for the waves of privatisation that spread outwards from Britain and America in the 1980s and 1990s. So we should know by now that what is sold as choice in this sort of Reaganite utopia often means the rich alone getting to choose and the poor making do with what is left.

*Angus Deaton, *The Great Escape: Health, Wealth, and the Origins of Inequality* (Princeton, NJ: Princeton University Press, 2013).

We might pay attention as well to what is left out. For as with almost all the mainstream approaches to development, this one too begins with a critique of one branch of economics, the better to replace it by another: in this case the insights of a younger generation of more energetic, field-oriented economists. This new generation of economists, perhaps best represented by Abhijit Banerjee and Esther Duflo in their influential work *Poor Economics*, enters the scene with a new instrument, the randomised control trial, slung across their shoulders like a bandoleer.* The RCT, in essence, provides comparative statistics to identify best practice across a range of contexts, and these new economists are determined to recommend only policies whose value can be *measured* on the basis of the RCT's results. If things can be measured, runs the logic, then they can be targeted, refined, and improved, and we shall have the success we desire without the cost or the political risk of needing to think so "big".

One can see the appeal. And theirs is certainly an approach that captures the spirit of our times. But is it really this simple? These economists seem convinced that they have put the ideology, not to mention the suits and cords, firmly to one side. And yet before they even make it into the field, measuring sticks in hand, there are plenty of claims being made: What is to be measured, and how? What values are we using to assess this? For all its claims to novelty, the RCT-based approach remains just another way of promoting the age-old fight *against* global poverty. Poverty remains the problem, not wealth, and the problem is still always located elsewhere.

Framed thus, and perhaps it seems useful to Banerjee and Duflo to inform their readers that "even people who are that poor are just like the rest of us in almost every way. . . . [W]e often find them putting much careful thought into their choices." Quite so. But surely this says more about the authors than anything else? More revealing still is their conclusion that the poor "have to be sophisticated economists just to survive." This is flattery to deceive: do the poor really get to count only when they are economic decision-makers, not husbands and mothers

*Abhijit V. Banerjee and Esther Duflo, *Poor Economics: Barefoot Hedge-Fund Managers, DIY Doctors and the Surprising Truth about Life on Less Than $1 a Day* (London: Penguin, 2012).

or carers or, dare we say it, politically engaged citizens? Banerjee and Duflo would likely point out that this is precisely the point: that it isn't their place to comment on things beyond their core competence as economists, that this is what turns many people against economics in the first place. But talking about world poverty is by definition talking about a great many things other than the economy: you simply cannot have it both ways.

The bottom line is that the turn away from the big and, as Banerjee and Duflo see it, academically uninteresting questions (the ones that are immune to empirical investigation of the sort that the authors are personally invested in), corresponds not only to a particular methodological mandate but also to a more deeply-seated tendency in our society to reject the idea of thinking in the round full-stop.

Yet the fact remains that there are some things we simply can't do without: health systems, for example. And for all the benefits of a targeted and local approach, a thousand impact-assessed health-care centres do not a health system make. Of course, evidence that aid is actually working wherever it is spent (including on health-care centres) is a benefit. But we should be wary of allowing only those things we can measure into the picture. For all the careful delineation of people's personal economic strategies in the RCTers "four cows, one donkey, and eight olive trees" approach to development, the needs of the poor as members of a community—let alone their need *for* community—all too frequently get lost.[19]

And all of this leads us back to square one: the fact that there is no single methodological fix for world poverty any more than there is a single economic fix for it (be it aid or anything else). And that means there must be other issues at play.

WHY NATIONS REALLY FAIL

Another branch of economics, the new institutional economics, does recognise the need to look at the bigger picture. Its adherents argue that the reason some countries are rich and some poor is ultimately down to a question of whether they have the right sort of institutions or not. Accordingly, politics is a matter of ensuring that poor countries adopt the right combination of institutions, chosen from a shopping list of specified desirable ones. Of course, to be desirable such institutions must derive from existing Anglo-American experience (they are institutions which, *post hoc, ergo propter hoc*, must be conducive to development). The shopping list therefore includes everything from electoral processes for adjudicating access to power (political power only, mind you: economic power is naturally assumed to be evenly distributed amongst all of us) to property rights systems for distributing rights to wealth, and on down the list to norms of governance, the law and even the black market (although in many countries, of course, the black market emerges precisely through people's efforts to *de*-institutionalise the economy).

A few issues are being swept under the carpet here. Property rights, to take what this literature itself considers a central example, may well be essential to countries' good economic housekeeping, as is claimed, but they are not always to be prioritised over everything else. Slavery was once a property right, after all, and the American market economy that these economists take as their *primus inter pares* (there is a lot of Latin in these books) was itself founded upon the destruction of the property rights of the king of England.[20] In any case, property is usually the one thing that poor people *don't* have—so a more exacting set of rules for what they can do with it is not of immediately obvious use to them. In short, there are times when, for ethical or practical reasons, developing nations may need to prioritise other things over creating the institutions that rich nations enjoy today. And it is not always nation-states, in any case, that should be doing the prioritising.

As with the RCTers we just encountered, there is also a very particular political ideology that grows like church ivy in and about all this methodological rigour. One influential and well-cited paper, for example, counts the idea of limited government (which is actually a value statement) as "an" institution while it chalks up the past thousand years of European history—hardly a golden age of peace and cooperation—as historical evidence that "good government" of this sort leads to economic growth.* Such generalisations are more sweeping still

*Rafael La Porta, Florencio Lopez-de-Silanes, Andrei Shleifer, and Robert Vishny, "The Quality of Government" (Working Paper No. 6727, National Bureau of Economic Research, Cambridge, MA, September 1998).

when the authors look beyond Europe. To wit: "We find that countries that are poor, close to the equator, ethnolinguistically heterogeneous, use French or socialist laws, or have high proportions of Catholics or Muslims exhibit inferior government performance".[21] We might wonder what they would make of Canada, then, in the near-Arctic north, with its French-speaking linguistic minority, its remnants of Napoleonic Code, its high proportion of Catholics (42% of the population), and its occasionally left-leaning government.

But why let an outlier disrupt a good theory? "The two central dangers that any society faces," write Simeon Djankov and colleagues in another famous paper, "are disorder and dictatorship."* Not poverty or disease or corruption, note, but that favourite conservative lament: anarchy and its discontents. So far as the "new institutional economics" is concerned, if you do not live in the Anglo-American liberal, democratic middle, then you either live in a failed state (society is falling into chaos and civil war about you) or you live in an autocratic one (you are toiling under the thumb of a feckless thug). The solution in both cases, it seems, is a form of order based on individual freedoms, which is to say, based on yet another value assertion: a state that is constrained by the market. By process of circular reasoning, we thus arrive at the justification we need for a liberal, free-market society that will be best positioned to provide these institutions of capitalism. And yet even here, we might note, the only sort of freedom really being promoted is that which comes from being a self-responsible economic subject.

A more convincing version of the institutionalist critique has been put forward of late by a number of authors, most recently—and influentially perhaps—by Daron Acemoglu and James Robinson, in their book *Why Nations Fail*. Acemoglu and Robinson hold Easterly and Sachs equally to task for focusing too much on the economics of underdevelopment and not enough on the politics. They claim instead that a country like

*Simeon Djankov, Edward L. Glaeser, Rafael La Porta, Florencio Lopez-de-Silanes, and Andrei Schleifer, "The New Comparative Economics" (Working Paper No. 9608, National Bureau of Economic Research, Cambridge, MA, April 2003), 6.

Egypt "is poor precisely because it has been ruled by a narrow elite that have organised society for their own benefit at the expense of the vast mass of people".[22]

This is true, of course, but by focusing only on the local, *domestic* politics of poor countries, these authors manage to again overlook the role of other political factors—in this case, of American aid in financing that Egyptian corruption, right up to the revolution. They applaud the effects of the Glorious Revolution in seventeenth-century Britain but deplore the idea of poor countries making their own revolutions today. The poor are thus criticised by Acemoglu and Robinson for the political decisions they were not in a position to take in the past and prevented from making the ones they are in a position to make today. We have indeed arrived at politics, at last, and writ large. But it is not an especially fair form of the art.

For all their talk of politics, Acemoglu and Robinson also sneak back in under the radar of this analysis a rather specific version of economics, since the history of institutional development is itself retold as a history of improvements and innovations made primarily by individual entrepreneurs like Thomas Edison. Either way, simply following the political recipe that rich countries established as they industrialised, but also plundered and fought their way to wealth, is not an option for poor countries today. As Cambridge economist Ha-Joon Chang argues, it is not just institutions that determine economic development; economic development also influences the character of institutions. Most of today's rich countries, he rightly points out, got the institutions said to be essential to the development of poor nations today only *after* they had themselves become rich. The Industrial Revolution thus came before the establishment of a banking system, of properly democratic governments, and of intellectual property rights—all of which are now pushed upon the poor world as drivers of development.[23]

But there are questions of power to be taken into account here too. As Marx once said, "Capital comes dripping from head to foot, from every pore, with blood and dirt." And the wealth of the British Empire was indeed built—as the historian Catherine Hall puts it—by the blood of sugar and slavery. Plantations provided much of the original capital the British used to invest in the Industrial Revolution, not to mention to build up markets (and protected ones at that) for their products. It was not, therefore, the technological breakthroughs of the Industrial Revolution that drove the accumulation of wealth and capital in Great Britain so much as the geography of empire, which allowed Britain to structure the international profit-loss balance sheet in its own favour.[24]

In fact, for an argument written to explain long-run change, once we have prodded and probed it just a little, Acemoglu and Robinson's world appears to be a curiously static one: wealth figures in the discussion, for example, but not the circulation of capital that creates it; meanwhile, human nature is unchanging over time. Elites are today, then, as elites ever were at the end of the Neolithic. Mankind is frail indeed. But can we really take no hope from the expansion of suffrage, the repeal of colonialism, the outlawing of slavery or the fact that it is racists, not racial minorities, who are more likely to be stigmatised today?

The history of the expansion of social welfare suggests we can. Indeed one of the first leaders to introduce social welfare policies was the man at the very top of the Prussian conservative elite of the nineteenth century: Otto von Bismarck. But this is not just about countering defunct models with better ones. It is about thinking progressively and finding what works.

THE TYRANNY OF GROWTH

For all the postured differences between the leading protagonists in the great aid debate and those who would usurp them, there is one thing on which almost all are agreed: the goal of development policy should, in the last analysis, be to bring economic growth to the poor world. But is economic growth really the elixir that they, or even we, require? In any case, is it best obtained through an economic model that its own champions struggle to construct a convincing historical model for? All things being equal, growth is an obvious thing to promote: how can it *but* contribute to human flourishing? But economic growth is good only if it is promoted as a means to prosperity (in the fuller sense of the term), not as an end in itself: as the only measure of prosperity.

All too often in today's world, social programmes are cast aside in favour of a narrow fixation on growth: this is true in rich and poor countries alike. But that fixation often ends up doing more harm than good:

the experience of India under liberalisation over the past thirty years is a case in point. Yes, India has been one of the fastest-growing economies in terms of growth in gross domestic product, averaging 6% a year. But the country continues to languish in 135th place on the UN Human Development Index, with 450 million Indians still living on less than $1.25 a day.[25] To be sure there are different types of "growth", and some are more pro-poor than others. But over the course of the past century, it is also clear that, all things considered, those who have pushed most strongly for growth have done so by mostly forgetting the most basic "condition of economic progress": that capital produces growth and not the other way round.[26]

We come now to the reason development's focus has always been on poverty and less so on inequality: for it is poverty that provides the discourse of growth with its normative power, and it is that power which makes development of use to the rich and not just the poor.[27] When growth is posed as the inverse state of poverty (and so presumed desirable), it becomes hard to resist anything that is not conducive to that growth. At the same time, however, the material qualities of poverty are enhanced (it becomes a condition to be ended or erased or cured, a trap to be climbed out of or otherwise escaped from) while its relational qualities (all the ways in which it is connected to wealth) are erased.

From this intellectual sleight of hand flow a good many of the most pervasive (and pejorative) stereotypes about the poor. Today we confront a mainstream developmental discourse that in some respects demands the existence of absolute poverty in order to be able to preach the party line of absolute growth as salvation. Yet it is actually far from clear that growth is the solution to the problems of an unevenly developed world—for all that this remains the predominant belief.[28] The World Bank long liked to claim—and most of us probably assume—that the greater part of the measurable decline in poverty and social well-being in recent decades is the result of economic growth. More people are better off if we have a bigger pot to share. This sounds reassuring. But the argument itself is a non sequitur: who knows what levels of poverty reduction might have been possible if things other than growth (redistribution, for example) had been prioritised? The far-too-uncritical celebration of everything that may lead to greater growth in today's mainstream economics ultimately detracts our attention from a single salient and underlying fact: Kuznetsian economics is a defence of structural violence.

There is much that is left out of the picture, then, in the focus on the donkeys and cows—or, at the other end of the scale, the GDP figures of economic growth. But suppressing this leaves us more than just intellectually impoverished. Because a politics of growth promotion that takes economic growth as an end in itself is actually the mainstream view in rich and poor countries alike: just get the economy going again, we are told, and everything will be alright. Yet this is a politics in which even well-intentioned efforts seeking to "overcome" poverty, all too easily drift into policies seeking to "tackle" the poor, as the bearers *of* that poverty.

In fact, no one has yet seriously shown that more growth without distribution of the gains is obviously better than less growth with distribution of the gains, especially when it comes to developing nations.[29] What has been shown is that the level of growth required to sustain the global economic system in the manner it is accustomed to working is itself unsustainable. Yet the development mainstream has closed its ears to this. "Growth is good for the poor", declared an influential World Bank paper in 2000—an assertion repeated and defended across the developmental mainstream.[30]

It seems it falls to a Marxist to pinpoint the problem here and a philosopher to point out that the problem was always obvious. To wit, David Harvey has long written about what he terms the 3% compound growth problem, where 3% compound growth is the minimum acceptable level of growth that a "healthy" capitalist society will accept, even as it is also a level that the market cannot absorb or sustain naturally (historical growth rates over the past 240 years average around 2.25%, for example). It may be useful to think of it as the Lance Armstrong theory of economic growth. The result is a recurring tendency to crisis, which hits the poor hardest of all.[31]

As the philosopher Michael Rowan points out, this doesn't even make sense for the wealthy over the medium term. However digitised or dematerialised our economy becomes, with 3.1% growth per year (as he defines it) we will soon run out of the resources to sustain that growth.[32]

Of course, economies respond to price and demand, and the economic reasoning behind some of the more fanciful estimates as to the environmental unsustainability of future economic growth is flawed. But however much we increase our productivity over sheer volume growth, the bar has already been set too high for everyone to take part sustainably.

A similar issue confronts the a presumed imperative of increasing all wages globally in line with what the rich already earn, since the level at which incomes and happiness diverge (at which point we stop getting noticeably happier with any additional unit increase in our income) is around $10,000 per person, and this is about the level of *average* global income already (if we could balance out the extremes). Above a certain level, growth in incomes and growth in the value of the economy at large do not equate to greater happiness, then, even for the wealthier among us. Moreover, any unit increase we may contrive to add to global economic growth is even less likely to benefit the poor, since historically it is those of us who are wealthier who mop up the benefits of economic growth *first* (where would be the fun in being middle class if that weren't the case?).[33] There are ways around this, of course. But if the price of sustaining our present levels of economic growth is that we all end up looking like China, then we have not succeeded in very much at all.

Contrary to the much-hyped benefits of economic growth to the vitality of nations, there is plenty of evidence suggesting that, over the longer term, it may in fact be the vitality of nations that helps foster the conditions for sustainable economic growth. The East Asian miracle, for example—all those tiger economies—is often invoked as proof that macroeconomic stability, competitive exchange rates, and openness to exports offer the best route to national economic growth. Yet it was frequently social norms and policies in these countries, themselves associated with declining inequality, that enabled the wage increases and savings pools through which these nations were able to take advantage of the benefits of more open economies in the first place.[34]

POVERTY TRAPS OR PROSPERITY PITFALLS?

There are two chief problems, then, with the debate on aid and development today. The first is that it is overwhelmingly a debate between different branches of economics, to an extent that excludes other, no-less-valuable perspectives. The second, related problem is that the entire debate has to date been monopolised by the assumption that the primary task of development should be to enhance economic growth (either of poor countries, in the macro version, or of poor people, in the micro version).* Both these tendencies need resisting if a new sort of approach is to become feasible.

There is a third problem underlying these first two, however, hardwiring them into the minds of the otherwise intelligent people who make decisions on all the above: a misunderstanding of the geography of poverty and inequality.† When experts talk about the "curse" of geography that afflicts poorer nations with natural resource endowments that lead to corruption, or when they claim that a nation's environment is the reason its people are poor—a poverty that oozes from their pores and sticks to their skin, as if there could never be poverty in Svalbard or Murmansk—they are implying both that poverty is the very property of the poor and that it is *located* in certain places. This is exactly how the urban poor of the Victorian age were once described. But the effects of this misunderstanding are today international in scope.

There is a corresponding tendency, for example, to reduce the complexity of people's lives to a recognisable variable, like corruption, which can then be projected back onto their local environment. When, in his highly influential article (and later book) "The Coming Anarchy", Robert Kaplan writes aghast of how, in Abidjan, "groups of young men with restless, scanning eyes surrounded my taxi, putting their hands all over the windows",‡ this one image of menace and threat is implicitly combined with others and collapsed into an explanatory variable applied to the rest of the country, if not the "Third World", at large. It as

*Erik S. Reinert, *How Rich Countries Got Rich . . . and Why Poor Countries Stay Poor* (London: Constable & Robinson, 2008).

†David Landes, *The Wealth and Poverty of Nations: Why Some Are So Rich and Some So Poor* (New York: W. W. Norton, 1998).

‡Robert Kaplan, "The Coming Anarchy: How Scarcity, Crime, Overpopulation, Tribalism, and Disease Are Rapidly Destroying the Social Fabric of Our Planet," *The Atlantic*, February 1, 1994.

if a street brawl after the football derby in Manchester were adequate to explain the corruption of the British lobby system.

In his book *The Bottom Billion* Paul Collier more elaborately introduces us to four basic (and highly influential) "poverty traps" that he believes explain the persistence of poverty among the bottom billion of the global population: each of these traps—resources, conflict, landlocked nation, and bad governance—works to locate the causes of poverty in the cohort groups that Collier has chosen to study. Collier uses various indices to quantify the features of these traps, although the fact that they fall back partly on studies drawn from 1980s Soviet ethnographies suggests the difficulties of trying to draw accurate statistical generalisations. He is nonetheless emboldened, as Banerjee and Duflo are, by what he takes to be the scientific rigour of his chosen methodology. By reading African reality back from his presumed objective framework, Collier conjures hypotheses from geographical assumption. Switzerland is a landlocked nation, after all, and Norway has resources. But Collier—and the World Bank that listens to him—is clear as to how the rich should be judged: "citizens of the rich world are not to blame for most of the problems of the bottom billion," he says. [35]

Yet as with Easterly, it seems not to be of interest to Collier that what looks to him like local bad governance may be imposed from without by international pressures, or that it may be the legacy of earlier distortions of the local political system inherited from the West at the end of a colonial era.[36] It is not so much geography that traps these individuals as history, then. And merely connecting them up to the global economy is no guaranteed solution either: as the Nobel Prize–winning economist Amartya Sen has pointed out, the poor are in fact most vulnerable shortly *after* they have been drawn into a wage-earning economy (and before they receive any protection from it).[37] The street corner of the free market, to which Collier tells us we merely need to give the poor a lift, is presently the very place where they are most likely to get clapped about the head and mugged.

But why not just clear the ground altogether and start from scratch? This was of course the instruction given to Baron Haussmann in nineteenth-century Paris, as he set about ripping away the alleys of the poor and making them over into boulevards for the rich. It is also the view today of New York University economist Paul Romer, a man on a mission to sell the idea of charter cities to poor countries like Madagascar and Honduras. Romer's idea is to establish what he calls "oases" (a revealing choice of word) of technocratic "sanity" (ditto), where, as one reviewer puts it, "struggling nations could attract investment and jobs; private capital would flood in and foreign aid would not be needed".[38] We are already familiar with this drama, of course, before the curtain even rises.

Such a business school approach carries with it the unwelcome echo of the politics of imperial urban design under the Raj. The countries where these cities will be established will be encouraged to "lease" land to foreign interests that will run their new cities for them, rather as the British did after seizing control of Hong Kong—an example Romer in fact looks to learn from.[39] Yet Romer and other poverty entrepreneurs—amongst them Sun Microsystems founder Vinod Khosla, who believes economic development is usefully thought of as a "chemical" process—seem untroubled by this.[40]

So too does Indian industrialist Ajit Gulabchand, head of Hindustan Construction Company and the man behind a planned cluster of semi-private cities to be built four hours from Mumbai, under the name of Lavasa. The first of these cities, Dasve, has already been built. It is modelled on Portofino, on the Italian Riviera. Designed to appeal to India's growing elite, Dasve "extracts the rich from Indian society into a European vision of prosperity", as Jason Miklian and Kristian Hoelscher write in the *Harvard International Review*.[41] Meanwhile, the construction and hospitality workers are housed in a separate settlement nearby. The likelihood of Dasve succeeding looks slim. But what matter? When things fail, neither Gulabchand nor Romer nor the foreign interests they court will be blamed, but the poor who failed to make things work.

But the problem here is not really the technical feasibility of such plans, or the fact that the rich like to have their little schemes; it is the relationships of power that are implied and imagined, and the anti-democratic tendencies that such schemes wittingly or otherwise seek to consolidate.[42] Questions of power are at the heart of every issue of the development of societies and nations, and above all they are at the heart of why prosperity comes to be located in some places, such as London, rather than others. Yet most mainstream theories about the world's poor suffer from a notably underdeveloped conception of power and where that power is to be found in international politics today: if anything these theories prefer to ignore questions of power altogether.

Glance at a newspaper today, and you'll see that what was once called "hunger" now goes by "failure of food security", for example. It is not immediately clear how this technicisation of the problem aids people who cannot produce enough food any longer because they have been forced to sell their land. But it *is* clear that it makes it easier for others to distance themselves from the far-reaching effects of their own patterns of food consumption. It makes it easier for us to all overlook how it is precisely in order to meet that demand that Western, and now also Chinese and Saudi, agribusiness is dispossessing people of their own land.

The problems we are encouraged to do something about in most mainstream approaches to global poverty today are, in short, such problems as are recognised by the wealthy and the powerful: and whatever the next fad in development thinking turns out to be, this is unlikely to change. Indeed, it cannot change until we start acting like the problem of uneven development is a political problem first and foremost, one that requires us to engage in a more democratic discussion about how rich and poor alike can work to reduce inequality in this world.

But before even that, we will need to be convinced: if our current way of thinking about the relationship between wealth and want is flawed, then what do we stand to gain by revisiting, a little more honestly, the way that wealth and poverty actually *have* developed together in the modern world?

THE POVERTY OF ELSEWHERE

3

The study of the causes of poverty is the study of how a large part of humanity is demeaned.
—Alfred Marshall, *The Affluent Society*

In 1948 the world pledged itself to an international bill of human rights. But somewhere between the ideological caprices of the Cold War powers and the market libertarianism of the post–Cold War era this promise has been squandered. Swords have not been turned into ploughshares; the meek have not risen up alongside the rich. For the past half century we have obsessed about the state of the world and fetishised the plight of its poorest people, but we have done so through the filter of a steadily diminishing recognition that our own choices are in any way what undermines the freedom of choice of others. This is what lies at the heart of the seeming intractability of global poverty. But we need history and politics—not just economic theories—to understand why.

By the late 1960s many of the countries in what was then called the Third World were veering onto just the sort of path that some say they now need to take again today. They had assembled themselves from the splinters of independence into a viable bloc with a collective agenda and a clear sense of what was required, by way of international diplomacy, to bring the poor into reckoning alongside the rich. By the 1980s, the global South—as the Third World was then called, as if that were a consolation prize—was ruined as a political force, while the global North, contra how we tend to remember things, was busy waging economic war against it. In many cases the North was helped in this by local leaders in the South intent on steering their nations into penury in exchange for their own personal gain. But we must not forget our own role in the making of an uneven global order.

The history of the "making" of the Third World since 1945 reveals two recurring tendencies. First, despite what the pessimists say, there is every reason to believe that poor countries *can* overcome their poverty when they are not exploited from without and when they are able to contain domestic exclusionary politics as well. Look back to the figure showing historical inequality rates in the twentieth century. It is clear how inequality was dropping for most of the world throughout the postwar period until the late 1970s. Even including Africa, inequality had been worsening only very slowly to that point. Second, it is frequently the rich world's need to solve its *own* problems that leads it to ensure that the global playing field remains skewed in its favour (the kink in both lines on that figure was caused, as we will see, as much by domestic considerations in rich nations as by anything in the poor world).

It is helpful to approach this history with the notion of the poverty of elsewhere in mind: think of this as a simple sociological tool, a reminder to pay greater attention to the ways that the rich distance themselves from the material consequences of poverty in the world. If we can do this, the actual politics behind the economic history become a little clearer. We have already seen how, by constantly reinventing the problem of poverty, the postwar generations made it easier for themselves to overlook the dynamics behind poverty creation and to "naturalise" the creation of wealth as mankind's raison d'être. But the quite substantial effort it takes to also keep wealth secure and in place is something we have learned to overlook. We need some way of correcting for this. We are accustomed to being told that we should be doing more to help the world's poor; first, we might do better asking what it is we have already done to put them there.

THE RIGHT KIND OF REVOLUTION[1]

During the Cold War, many people—Western scholars in particular—looked to what they called the Third World as a source of hope. It is more common today to look to the poorer countries with anxiety: in fear that they are taking our jobs from us, or harbouring the next generation of terrorists. Such anxieties may be misplaced, but they will retain their grip on the Western mind until there is more widespread recognition of what really has happened in and to those countries. As the Argentine philosopher Enrique Dussel reminds us, the quite distinct histories of the rich and poor world really make sense only when seen together.[2]

In the aftermath of the Second World War, a belief emerged in the rich countries that a better world could be had by taking the lessons of the West and applying them to the rest. With new planning techniques and statistical science at their fingertips, the brightest minds believed that they could shape the world to the best of all possible ends. After 1945 the technical and utopian aspects of this contention merged into a seamless whole: development. Yet the outlines of a subsequent betrayal could already be detected in the assumptions that our past was also their future and amidst the biological metaphors that not infrequently sought to explain this "natural" ordering of societies.

It seemed, then, that the late nineteenth-century vices of racial planning and social engineering—banished from Europe, where they had wrought such damage—were acceptable still when applied to the margins of the world. For all the aspirations on paper for a postwar era of human rights and equality, there was always a reluctance—in part a hangover from imperialism, in part simple prejudice—to accept that these things really might apply to the world's poor. The rich world hesitated to give up on the belief that people in poor countries were not to be trusted with quite the same freedoms the rich already enjoyed, and for all that colonialism was "ending", other forms of exploitation—"special" trade agreements and first-refusal clauses—were just beginning. Cold War politicking soon limited the very question of freedom to matters of ideology, but we managed to notch up even this sleight of hand as a case of moral virtue.

Compared to the arms race, development policy may well have seemed a more enlightened Cold War front, then. But it was no less ideological. Walt W. Rostow, author of the influential *The Stages of Economic Growth* and adviser to President Kennedy, always believed that development policy was a weapon: an ideology best deployed "crabwise", as he put it, through the technicalities of modernisation theory.

After all, as is often pointed out to undergraduates, the subtitle of Rostow's great work was "An Anti-Communist Manifesto".[3]

The presumed lessons of Europe's own experience were important to post-1945 development in another way too. The European Recovery Programme (more popularly known as the Marshall Plan) of 1948 offered President Truman a clear demonstration of the benefits of foreign aid to Americans and American foreign policy. It helped to convince him, and the United States, of the value in taking an interest in the internal affairs of other countries, and eventually to tack on to the postwar American foreign policy agenda, the so-called Point IV Programme of 1949, a commitment to developing countries south of the Mediterranean.

If Truman hoped this would steal a march on the Soviet Union's own extra-European ambitions, he was right. Joseph Stalin was never much interested in the world beyond the Russian heartland, and the days when the Comintern spoke meaningfully of international solidarity were long past. But Nikita Khrushchev, Stalin's successor, was much happier to take up the development challenge, as he made clear during a tour of Afghanistan, India, and Burma in 1953. "Perhaps you would like to compete with us in establishing friendship with the Indians?" he crowed at the Americans a couple of years later. "Let us compete," he declared.[4]

A farmer himself, Khrushchev was well aware of the Soviets' unique selling point, as we might put it today, relative to the more immediate attractions of American-style mass Fordism: the Soviet Union had itself recently been poor, so it could portray itself as having a natural affinity with the Third World. And on this, he was right. The reason many poor countries, upon achieving independence, actually chose to follow the socialist rather than the capitalist path had much more to do with their need to address acute agricultural and industrial underdevelopment than with any desire to bind themselves hand and foot to Marxist-Leninist dogma. It also had to do with their pride at finally having this thing called a state that could be put to work to achieve their ends (though not always beneficent ones, it should be said).

The state was also a crucial part of the American variant of modernisation in those early years of development. In the initial postwar years, the intellectuals who staffed what were, after all, usually American-endowed international institutions were primarily the New Deal warriors of yesteryear. But a new generation of administrators who had taken over the reins by the beginning of the 1960s were much more sceptical of the whole philosophy of New Dealism and they soon set about dismantling it.[5] With the spectre of communism lurking, preaching the power of the state was said to be a dangerous idea.

America's retreat from its version of the state as development tool was, in many countries, the Soviet Union's gain: since many newly independent countries took sanctuary instead in the Soviets' core offering of the five-year plan. The idea of five-year plans seems a laughable anachronism today. But as it happens, they are alive and well in most international and non-governmental organisations. Perhaps, then, it is merely the idea that *states* should plan that we find ridiculous. In any case, the planning approach was welcomed in the mid-twentieth century as "a bright and heartening phenomenon in a dark and dismal world", as a young Jawaharlal Nehru recalled of his first visit to Moscow in 1927.[6] He was not alone. When in 1962 someone had the smart idea of asking African students in Paris which superpower they preferred, 25% opted for the Soviet Union. Only 3% chose the United States.[7]

Such figures help explain why in the United States it was felt to be such a strategic necessity that the American vision of development be able to point to rapid material progress. If it were to do without the state to distribute resources and to build the institutions of social progress, then *this* version of development would have to be able to leverage the power of markets as a way of creating rapid *material* progress first. The West needed to make the benefits of American values as apparent as possible to the peoples of the Non-Aligned world: the great global swing state of the mid- to late twentieth century. And this meant that Western-style development became increasingly synonymous with a single measure of prosperity: economic growth.

We live, as we have seen, in the shadow of this (ideologically driven) theory of economic development still today. But we live in the shadow of other ideas from the past too. In the 1950s and 1960s US experts were keen to frame poverty as the symptom of specific, locally rooted problems such as hunger and disease, problems for which American technological supremacy could then be presented as the cure (a lesson Bill Gates has learned well).

To the problem of hunger, for example, the West offered the Green Revolution, which secured a Nobel Prize for Norman Borlaug, the agronomist behind the high-yield seed variety it was based upon. But the Green Revolution was always as well a quick fix for two increasing Western concerns. First was the "population bomb" feared to be awaiting poor countries and, by extension, the world—an anxiety reflected in Garrett Hardin's famous 1968 "The Tragedy of the Commons".[8] Second was the need to exert some measure of control over what was taken to be the alarmingly freewheeling politics of recently independent poor nations. Who knew what they were going to do next? Better to tie local farmers into foreign crops and fertilisers provided by the state, and then further tie those states into the West's agricultural company ledgers and aid programmes. When it comes to the small print, the free market was indeed never quite so free as it appeared.[9]

The communists made even grander promises than this, frequently with even more dangerous small print. Socialist theory as it was taken up in the domestic vernacular of writers and leaders like Frantz Fanon, Che Guevara, and Mengitsu Haile Mariam, sought a "new man" to rise to the challenge of a brave new world. But for the citizens of many countries it wasn't always clear that their new saviours were on their side. For them, the violence inherent to the dangers of dreaming of a state where "tout est ordre et beauté!", as the economist Albert Hirschman warned, remained apparent.[10] Perhaps the central case in point here is the thoroughly mixed history of African socialism, which in the end came to encompass everything from Gamal Abdel Nasser's Al-Mithaq (Socialist Charter) of (ultimately unaffordable) health, education, and housing provision to Milton Obote's Common Man's Charter of 1967.[11] For a good many of the peasants selected to benefit from such schemes, "being opposed to 'progress', whether it was the *grands projets* of the colonisers or the collectivist fantasies of the new elites, was dangerous and usually deadly", as one historian has put it.[12]

For all their efforts to find their own feet, poor nations were hard pushed to avoid the visions of development of one or the other side in the Cold War. For every socialist Ujamaa village in Tanzania there was a capitalist model hamlet; for every "new man" sowing corn in his downtime from educational duty, there was a utility-maximising farmer entering into the market to buy modified Western seeds. Yet for the Third World, the really tragic irony in all of this was that the emergence of détente between the United States and the Soviet Union in the 1970s would make things worse, not better.[13] As the surefire assumptions of America's "all in" approach to development—"modernisation" in Chile, "Food for Peace" in India, high-yield crops in Mexico and the Philippines—vanished into the ashes of Vietnam (along with the careers of committed stalwarts like Walt Rostow), the flow of funds to the Third World dried up. As far as the Cold War went, poor countries were damned if they did and damned if they didn't.

THE ROAD NOT TAKEN

When looking through the tinted glass of "failed state" theorists and neoclassical economics, it is easy to look to the poor world today and point to what it has failed to achieve. Yet when seen in light of the challenges it inherited at the close of the Second World War, one finds reason to be impressed that it has achieved much of anything at all. In 1947, even India—the jewel in the crown that had long been preparing to go its own way—had, at the time of its independence, yet to integrate the princely states. It also had yet to build up a class of bureaucrats, it was dealing with an unprecedented population explosion (though it didn't know it at the time), and it had just "loaned" to Britain, via forced credits collected during the war, a sizeable portion of what little money it had to start out with as a newly independent nation.

But at the height of the 1950s and 1960s, as development and decolonisation proceeded alongside each other, the modernisation theorists in the West were joined in their enthusiasms, if nothing else, by the first generation of leaders of the newly independent nations. As a result, many countries wound up caught between the unrealistic promises of their leaders and the inbuilt weaknesses of their inherited, post-colonial states. One is reminded of Chinua Achebe's retort to Kwame Nkrumah's famous injunction ("Seek ye first the political kingdom, and all else shall be added unto you", Nkrumah had said to his compatriots): Well,

Achebe wrote, "We sought the 'political kingdom', and nothing has been added unto us; a lot has been taken away."[14]

But for all that we remember the past of poor countries as prone to ideological manias and transfixions, the majority of them in fact followed a distinctly pragmatic path. Indonesia's five-year plans were little more than "ornaments of state", wrote the American ambassador there in 1961.[15] But the enactment of constitutions with civil liberties guarantees and adult suffrage, or ministries built in concrete and steel, in which a national intellectual class set to work pushing paper and plotting graphs was no small achievement at all. Then as now Fidel Castro was an arch-ideologue to many—"to create wealth with social conscience" indeed. Yet the actual work of Cuba's Central Planning Board adhered most consistently to a policy of learning by doing.[16] And it usually found ways of working around the worst of the hubris in Castro's speeches.

More so than the Cubans, India's Nehru was decidedly a pragmatist in his policy prescriptions. Indeed, Guevara would be singularly disappointed, from a revolutionary point of view, when he met Nehru in 1961. One could not worry about distribution, Nehru once said, until one had something to distribute. He was also a democrat by instinct. To wit, Nehru kept the Soviet-style format of his beloved five-year plans, but he sent his engineers to train in Massachusetts. But then what choice did he really have? A global market economy already existed, and the task was to forge ahead within it. As he and every other leader across the poor world was only too aware: "laissez-faire was planned; planning was not", as Karl Polanyi insightfully summed it up.[17]

The poor world's leaders all had to deal with this. Contrary to popular myth, the way in which most of them did was less radical than we recall today (with one or two colourful exceptions) or than they professed it to be at the time. Populists were invariably forced to accommodate nationalists, while nationalists in turn were more diverse than their anti-colonial rhetoric suggested. The majority of the new governments also recognised that unemployment was not the major concern of most nations at independence, unlike for Western economies. In confronting, rather, the "unholy trinity" of "poverty, ignorance and disease", it was health and education that needed to come first.[18]

In countries where this was most strongly recognised the result was a commitment to *social* policy as an integral part of economic planning, and this came much earlier on in these countries' economic development than it did in the West. Many such initiatives, like the great era of the five-year plan itself, were optimistic, to say the least. But even some of the more grandiose official statements coming out of countries such as India—about "welfare states" and "cooperative commonwealth"[19]—were a reminder that such rhetoric was at least presumed likely to capture the public imagination. Even as late as the 1960s big business in India routinely used the word "socialism" in its own literature without considering that doing so was in any way unusual.

Though partial, such welfarist gestures were hugely important given the background disorder they were frequently posed against—this was also the era of Korea, Vietnam, and a wave of revolutions across Africa and Latin America. They pointed to what was possible when political will existed. And they suggested—quite against the one-size-fits-all claims of Western or Soviet strategies for modernisation—that a variety of more integrationist social policy approaches could be undertaken in a variety of different contexts, and most would probably work.

Indeed, these social policy programmes had a demonstrable effect. By the late 1970s, when these programmes were at their height, per capita income of "underdeveloped" countries was growing by close to 3% (and this despite population growth of 2%), along with increasing real incomes and declining child mortality rates (suggesting the growth wasn't just being gobbled up by the rich). Average growth for sub-Saharan Africa in particular topped 4.3% for the period 1967–1980, what some refer to as the continent's own "golden age". Between 1945 and 1981, the eight largest countries of Latin America all posted growth rates exceeding that of the United States. And nine of the world's thirty fastest-growing economies were African, and nine Latin American.[20]

AN ERA OF BENIGN INTERNATIONALISM

All of these achievement occurred largely despite Cold War politicking and, amongst other things, the Central Intelligence Agency's particular brand of policy-making by the knife. But it was also achieved with the help of a more amenable international economic environment than we have today: backstopped mainly by the United States, it should be said. This is frequently forgotten, not least by those who wish to insist that no usable feather has ever fallen from old Uncle Sam's cap. The Bretton Woods era is remembered for having strongly promoted free trade. It also insisted upon strong regulation, and strong capital controls in particular, "not merely as a feature of transition, but as a permanent arrangement", as Keynes put it.[21] But this aspect of the era is less often mentioned. If we had remembered just that, we would not be in quite such a dire situation today.

These capital controls were not introduced out of altruism, however, but more often out of concern for national and international security. Yet as the political scientist John Ruggie convincingly showed some time ago, the controls nonetheless actively encouraged an "embedded liberal" order in which "the policy autonomy of the new interventionist welfare state" was to be protected, and not just at home.[22] In short, just as national elites may tolerate for a while things that are not in their immediate interests but that they may benefit from over the long run, so too did the international community once act in just the same way.

Today, it is increasingly clear, we need to get something like an *embedded* liberal order back, and the reasons why are worth reviewing. America's postwar commitment to the values of embedded liberalism—that is, a liberal political order that accepts regulations on economic activity—had the distinct merit of seeing the market as the servant of the state. This was not because peoples' beliefs were so very different in those days: people were no less entrepreneurial; no less fascinated by the latest technological breakthrough than we are in the age of the smartphone. Indeed, a more "embedded" approach to the economy had to be actively fought for against those who said to leave everything to the market to decide, and in opposition to the sort of policies that the era's own monetarists and free marketeers (men like "money doctor" Edwin Kemmerer) had pushed both at home and abroad before the Second World War: policies to which we have today largely reverted.[23]

But during the relatively short-lived sway of Keynesian internationalism between these two monetarist epochs, the United States, as the international political economist Eric Helleiner shows, actually encouraged poor countries to exert some measure of control over the flows of money and goods across their borders, via the setting up of national banks in countries like Paraguay, Honduras, and the Philippines that were responsive to domestic economic needs. In their own spheres of influence Britain and France by contrast sought to retain the old colonial currency boards, or to otherwise limit national autonomy. When today the United States wants to point to the good it has done, it should point, if anything, to these policies, not to the fluff of so much "democracy promotion" that has coloured its every policy since.

But once again, these policies were only partly about altruism. They were at least as much about foresight. By giving individual nations the ability to respond to external economic shocks, the wider international system would be strengthened and become more durable.[24] The approach worked for either reason. And the plethora of concrete prefab construction that continues to greet visitors to the developing world is testament still to what was then achieved.

By the mid-1960s the appeal of America's more open-minded approach to the world economy appeared to be catching. Even Britain could be seen looking with favour upon its economic relations with the Commonwealth nations and not just with the OECD: "We are not entitled to sell our friends and kinsmen down the river," declared Harold Wilson in 1965, "for a problematical and marginal advantage of selling washing machines in Dusseldorf."[25] This was, of course, as much about Britain's own brand of Euro-scepticism as magnanimity towards its former colonies. But the European Economic Community itself soon approved the Generalised System of Preferences, allowing tariff-free imports of manufactured goods from poorer countries into the area: a move welcomed by the poor countries as a "revolutionary step" for the way it seemed to thumb its nose at the General Agreement on Tariffs and Trade's free-market agenda.[26]

And with the Lomé Convention of 1975 (a trade agreement approved by all the Western European trade unions to extend preferential access to the European Economic Community to a full forty-six African, Pacific, and Caribbean signatory countries), the European Union broke new ground by providing for aid transfers the other way. This was followed in 1976, under the aegis of Valéry Giscard d'Estaing, by the

Conference on International Economic Cooperation, through which Europe hoped ultimately to secure stable oil prices for the north, in exchange for which it would consider the wider developmental needs of the commodity-producing nations themselves.[27]

To be sure, the willingness to recognise the specific needs of poorer countries had its limits (though compared to the present-day system of free-trade agreements that the European Union is lining up with countries like India, it looks fair by comparison). And one could draw those limits in any number of ways: Vietnam, Suez, El Salvador. But it saw at least a certain space open up for the poor world to itself contribute to the emergent postwar international order. And those contributions proved of value to the rich world too. When the US Federal Reserve began to question orthodox monetarism during the war, it was in part because Latin America's abandonment of the gold standard had already inaugurated a successful period of activist economic policy, which helped re-establish economic growth after the Depression in the United States' own hemisphere.[28] Likewise, Lyndon Johnson's War on Poverty in America, which began with the Economic Opportunity Act (1964), was influenced by the anti-poverty programmes that had been undertaken in developing countries.[29]

THE RISE AND FALL OF THE THIRD WORLD

We deceive ourselves, of course, if we eulogise what was sometimes little more than a means of finding novel ways for the "new" countries to contribute to the coffers and the standing of the old. In Europe a colonial mind-set persisted, and the habits of imperial orchestration died hard. In America too, J. Edgar Hoover insisted on seeing "a Communist behind every tree"—which shows just what America had learned from Europe about denigrating ideas by reference to pre-existing prejudices.[30] It was always going to take more than a mildly altruistic Western pragmatism, therefore, to keep ajar the window of opportunity that the postwar order of embedded liberalism had opened.

What it took, in fact, was the concerted pressure of the Third World itself, acting together as a political bloc, for the first time in history, as it fought for its own survival. In 1955, the dominant figures of the Third World met in Bandung, Indonesia, to establish the terms of this political bloc. The Non-Aligned Movement was formerly inaugurated seven years later in Belgrade in 1962, and via this movement's influence on the floor of the UN General Assembly, as well as through its own meetings, the poor world was able to speak with its own voice within a post-1945 consensus still largely determined by the rich world.

But as its membership expanded, the Non-Aligned Movement struggled to retain the cohesiveness, let alone the progressivism, of its earlier years. Its member states suffered too after the passing away of some of the original personalities who had held "the project" together. Nehru died in 1964, and his rather tepid successor, Lal Bahadur Shastri, immediately set about liberalising markets at home while withdrawing India from the movement.[31] Indira Gandhi in turn was a more substantial leader, but one whose policies flip-flopped interminably. Meanwhile, Sukarno, the primary figure behind the meeting at Bandung, was ousted by the authoritarian Suharto, and Anwar Sadat, Nasser's anointed successor, was assassinated in 1981 (although not until he had unceremoniously ditched most traces of Nasser's socialism). And of the new generation who took over, including the likes of Hosni Mubarak in Egypt, most would prove more interested in playing their countries' long-term interests off against their own personal short-term gain.

Perhaps more important, the original and often quite effective consensus policies of the Non-Aligned Movement were increasingly sidelined by a more impatient wave of insurgencies and guerrilla struggles that today we take for the entirety of the Third World's contribution to politics. These struggles flared up during the 1950s and 1960s everywhere from British Malaya to French Indochina, to Cuba, Guinea, the Congo, Bolivia, and of course, Vietnam. There was never any special logic to the timing of these struggles, with each largely a response to local conditions. But they were all in at least some measure born of a similar frustration with the increasingly unhelpful combination of Western intransigence, domestic failings, and the Realpolitik of the Cold War.

The revolutionary wave of Third Worldism reached its greatest intensity in the late 1960s, its zenith coming at the Tricontinental Conference held in 1967 in Havana, where banners wafted in the sea breeze and declarations were shouted out into the tropical heat. But it had run out of momentum by the start of the 1970s, no less than had Che Guevara, who radioed in his last public message to the Tricontinental from his jungle hideout in Bolivia. The Third World was then left with little more to assert itself than the price at which it sold its commodities: to wit, the Organisation of the Petroleum Exporting Countries hiked prices twice, in 1973 and 1978.

By then the West's period of benign hegemony was itself nearing an end, with the move from the gold standard to floating exchange rates in 1971. For the poor countries this meant that their already-poor terms of trade worsened, and their development flagged just as the rich world began losing interest in foreign aid as an arm of foreign policy. By the middle of the decade, a number of countries were already in economic trouble, and by 1981 none of the poorer countries showed any growth at all: a first in more than a quarter century.[32]

In short, whatever doors may once have been open to poor countries were now quite firmly closed. Revolution had failed, the openings to Western trade policies were shut off and sealed, and there was increasingly only one route left: accommodation and adjustment to the economic needs of global capital. The entire world economy was in the process of rearranging itself from an increasingly vibrant bazaar into an auditions queue for Western industrial manufacturing.

But even here, China, which was about to make itself a more attractive destination for investment by offering lower manufacturing wages than any other country was able to, was already preparing a trap for those who took the low-cost labour route. The Trilateral Commission of American policy makers who first met in 1973 can thus, in hindsight, be taken as something of a riposte to the triumphalism of the Third World's Tricontinental Conference, just six years before. But that commission also initiated a new round of international *raison d'état* and a new era in the relationship between rich and poor in which Western norms and procedures, and above all a conflation of the military and the economic, would be ever more strongly advocated for: until they were literally pressed into the emergent institutional fabric.[33]

"ORGANISED SABOTAGE"

By the 1970s the Third World was all too aware that the era of embedded liberalism had passed, and it had begun to focus its energies accordingly on the call for a New International Economic Order. All too often overlooked today, the NIEO—as it was tabled within the United Nations in May 1974—was a remarkable moment of political possibility, calling for "equity, sovereign equality, interdependence, common interest and cooperation among all States, irrespective of their economic and social systems which shall correct inequalities and redress existing injustices". With hindsight all this was perhaps a little *too* remarkable: for if the impact of the NIEO owed much to the first OPEC price hike, so too did the tenor of the response to it. Nonetheless, history ought to record that this was the vision of the world that the developing countries called for in democratic voice. It was not the one they got.

What they got was a new world order of the rich world's making.[34] Reeling still from the first OPEC price hike of 1973, and with the military threat of the Soviet Union receding, the prime concern of intellectuals and policy analysts in Foggy Bottom now became the economic threat of a Third World ascendant. Fearful of the power of OPEC and enraged at the temerity of the NIEO, men like Daniel Patrick Moynihan, former US ambassador to the United Nations, spoke openly of "Third World Extremism",[35] as if geography itself were a form of terrorism, and they began to plot the West's response accordingly.

The response took two forms. It began, first of all, with kicking the United Nations itself—where the Third World had most real leverage—into the long grass. As the historian Mark Mazower shows in his reconstruction of this period, the United States sought to convert the UN General Assembly into little more than a glorified talking shop while evacuating real decision-making power to the Security Council. At the same time the UN Conference on Trade and Development—the Third World's main hope for reform of the international trading system—was "put on ice"[36] and the work of independent bodies like the International Labour Organisation was pushed into the shadow of the more powerful and, for American interests, more amenable institutions of the Washington Consensus. Thus the international order as we know it today was formed: an order of selective and strategic reorganisation—or "organised sabotage", as the Swedish economist Gunnar Myrdal put it.[37] The results of that sabotage were soon clear for all to see: in the 1950s and 1960s, a government defaulted on its debts on average once every four years; from the 1970s onwards, that rose to a rate of four every year.[38]

The second blow to the Third World's hopes that it might, through its own good works, come to stand alongside the rich nations in the court of global power came in 1975, in the sleepy suburbs of Paris. There, at Rambouillet, the traditional summer residence of the French president, what was then the G6 club of wealthiest nations—America, Britain, France, West Germany, Italy, and Japan—was founded and a new era of international decision-making ushered in. The world was in the deepest recession since the 1930s, and since none of the key figures behind the meeting—neither Kissinger, Kohl, nor Giscard d'Estaing—had any clear way to address this, it was decided then, and it remains the working consensus today, that having the powerful and the like-minded club together was the next best thing to an answer. The G6 was thus yet another step away from the Roman forum of the United Nations General Assembly and towards the politics of the Privy Council that have operated ever since in the G6 (now the G8) and at the Security Council.

Between 1979 and 1981 what remaining chance there may have been for a more equal world was finally finished off. The Third World's two anni horribiles began with the "Saturday-night special" of October 1979. The initial interest-rate hike orchestrated by US Treasury secretary Paul Volcker on the night of Saturday, October 6, 1979, in an effort to control domestic inflation at any cost, was an example of what could now be done—fifty years after the Depression—with the new exchange-rate flexibilities that Richard Nixon had secured in 1971.

Still, having a gun is one thing; using it is quite another. And in economic terms, Volcker had gone nuclear. Three decades later, the wastelands of Philadelphia and Baltimore bear witness still to the destruction he unleashed. By deliberately and suddenly raising the US interest rate to what would ultimately scale 20%, Volcker plunged not just the United States but also much of the rest of the world into recession and unemployment.[39]

The consequences were soon apparent. The capital accounts of developing nations were $85 million in surplus in 1980; by 1984 they were $233 billion in the red.[40] These countries found themselves pushed towards a growing burden of debt, the value of which, being mostly denominated in dollars, was for the same reasons also appreciating rapidly.[41] For those Latin American countries coming off the more generous longer-term loans of the early 1970s and onto the rather more precarious short-term loans (all they could then hope for), the result would be near catastrophic.

Not surprisingly, it was these countries—first Mexico, then Brazil later that year, followed by Venezuela, Chile, Peru, and Ecuador the next, that were the first to default on their loans, inaugurating what has come to be known as the "lost decade" of development: a decade of lost lives and lost opportunities. The timing was cruel fate indeed. "Never before have so many people had so many life chances," reflected Oxford University political scientist Ralf Dahrendorf on still–social democratic Europe in 1979.[42]

For the rest of the world, *Time* magazine's Jay Palmer was closer to the truth: "Never in history have so many nations owed so much money with so little promise of repayment."[43] Most galling of all, many of those countries that were hardest hit were the very ones whose debt ratios were highest precisely because, like Europe and America before them, they had used large sums of money to invest in longer-term social policy.[44]

In the years since, the rich world's establishment has come to view the banking and debt crisis of the 1980s with something approaching relief: a bad turn, but a case of depression avoided. The poor world did not get away nearly so lightly: there was runaway inflation in Brazil (242.2% in 1985) and Bolivia (more than 8,000% the same year),[45] and across sub-Saharan Africa 134 million people were pushed below the poverty line between 1981 and 2001.[46] Yet the impact would be measured in more than money alone. Educational enrolment declined in many African countries during these years, with public spending on education in Nigeria, for example, cut back from 6% in 1980 to 0.6% as part of the country's structural adjustment programme.[47]

Having first tipped the poor world through the thin ice of overexposure to debt, the West then made clear, two years later at a pivotal meeting between North and South at Cancún, Mexico, in 1981, that the sort of politics in play was deaf to the sounds of drowning. When Indira Gandhi, in one of her more progressive moments, raised the idea of agricultural subsidies for India, Ronald Reagan replied, straight-faced: "This is cheating!"[48]

The billions poured into the US cotton sector and European Common Agricultural Policy must have eluded Reagan's recall. But Reagan's bullishness at Cancún was just the beginning. The years of containment and embedded liberal compromise had come to be seen as years of profligate waste, in which whole parts of the world had been needlessly ceded to communism and its discontents.[49] To put a stop to this, the United States made sure that it also poured money into arming or supporting rebels prepared to take a firmer stand against it, as they did in Afghanistan, Nicaragua, Cambodia, Angola, and before long, Mozambique.[50]

THE REBIRTH OF LAISSEZ-FAIRE

The real damage, however, was done in the late 1970s and into the early 1980s not through military or political aggression but through economic policy, as what began as a crisis of declining profits in the North (the problem of stagflation which Reagan and Margaret Thatcher had variously attacked) was steadily but relentlessly turned into a debt crisis located in the South. Net inflows of investment to poorer countries were soon dwarfed by even larger flows of debt payment coming out of the poor countries, and entire countries were cowed back into their old commodity-exporting role by the newly empowered institutions of the Washington Consensus. It seemed not to matter that during this time non-oil commodity prices were low and even oil prices were beginning to drop.

In short, by the mid-1980s the Third World as a whole was paying a very high price for the effrontery of OPEC a decade before. There was a wider logic to all of this, however, that remains with us today. For as the baby-boomer generation came of age, there was a desire in the rich countries to loosen the straitjacket of government regulation and the trade restrictions inherited from the 1930s. The "protectionism" of regulatory states was the wrong strategy for a globalising era, it was claimed—in a rather short-sighted reading of the economic measures first devised to mop up the crisis of the Great Depression (which, of course, was caused by the very unfettered monetarism now being rolled out once again).

To be sure, there was a lot that *was* problematic with the Keynesian state and the economic cataclysms of the 1970s had been a rude awakening to this on many fronts. But the response in the rich nations was not to try to correct for these, but to reach for the exact opposite: a desire whose doctrine was ready and waiting in the libertarian free-market philosophy of economists like Milton Friedman and Friedrich Hayek.[51] Having first lost faith in the Lutheran church of embedded liberalism, the rich nations, now firmly under the grip of the economic Counter-Reformers, set about vigorously depriving the rest of the world of their faith in the human powers of the state over the purity of market mechanism. The result was a new era of conservative internationalism: the gold standard was finished, Bretton Woods came to an end, and the ground was prepared for a new era of international governance, organised around the Washington Consensus: a more conservative at the World Bank, a more powerful IMF, and in due course a free trade orthodoxy institutionally embedded at the World Trade Organisation.[52]

There is much that can be said about the kind of globalisation that the new conservative internationalism helped to produce. But what is most worth bearing in mind is that at the same historical juncture that a "winner takes all" economy was successfully established in the North, a vocabulary of "basic needs" first began to appear in referring to the poor world.

This is not an innocent conjuncture and it reminds us, or it ought to, that the world's poor do not work for less than we do by choice. They do so in no small measure because of the situation they were put in around 1980, when credit-fuelled economic growth in the Anglo-American world fed off both the national debt crises in Asia, Africa, and Latin America and the need for individual credit in the rich countries.[53] This, of course, was one reason Britain and America were both at the forefront of lobbying for deregulation of the financial sector in those years (even as they were simultaneously lobbying for protectionist tariffs for their own textile and agricultural producers).

It is remarkable with hindsight that so much flawed policy could be peddled as so much incontrovertible doctrine. But the neoliberalist philosophy that stood behind this—a philosophy conceived as the property-owning class's response to the crisis of the 1970s—has always been more concerned with results than with consistency. And the trick of reversing the verities of cause and effect was equally a part of the neoliberal art then as it was in the 2000s, when government bailouts saved many an executive's hide, only by passing the costs on to the taxpayers. "What the world's poor countries need most is less government," opined the *Economist* in 1989 as evidence piled up of the failure to exercise sufficient economic oversight.[54]

And what the world's poor countries "needed", they got, as the neoliberal consensus of free-market economics, limited regulation, and "public choice theory" seeped into the cracks and whorls of the postwar international architecture and began to split it apart. Regulations were torn away so that rich and poor could trade with each other "equally"; a strict intellectual property agenda was pushed through in the General Agreement on Tariffs and Trade, ensuring that the spoils of the information technology revolution stayed with the rich world; and the poor world's labour markets were exposed, via de-unionisation in the West and free-trade agreements between rich and poor countries, to the full force of the principle of corporate shareholder return.

For poor countries, neoliberal policies meant governments indulging Western conservatives' suspicion of radical nationalisms by replacing "national development" with sectoral development and targeted reforms, in effect ensuring that only those who were most able and willing to play the West's game would see any benefit from globalisation at all. These policies also meant substituting non-governmental organisations (both domestic and international) for core functions of the state, further weakening government capacity in favour of governance strategies. They meant circumventing the voice of poor countries' citizens in frequently unfair and almost always undemocratic "consultations" between the international financial organisations and those countries' increasingly unelected representatives. They meant reinterpreting, as in colonial times, the waste of capital and resource that Western intervention was creating as evidence of those societies' inherent inability to take care of themselves. And they meant, above all, the "permanent discipline" of structural adjustment.[55]

THE FAILURE OF THE LEFT

What did the left, which had begun the century as an influential source of internationalist thinking, have to say to all this? In some ways a very great deal; in others not much at all. The year 1968, when students protested in cities across the West—variously (and variably) inspired by the likes of Guevara, Fanon, and Mao—appears today to have been the high point of poor countries' influence on European and American thought. It stands out too as the beginning of the end of the European (and what remained of the American) left as an effective force in national politics. It was nothing if not appropriate, then, that 1968 also saw the release, to widespread acclaim, of the Cuban Tomás Gutiérrez's film *Memories of Underdevelopment*: a story in which the protagonist—a bourgeois writer, critical both of his family members who had fled the revolution and of "the naivety of those who believe that everything can suddenly be changed"[56]—struggle to come to terms with the changing order of the day.

This was precisely the problem that was confounding the left at large. The revolutionary left—which had never really come to terms with the failure of either *der real existierende Sozialismus* in the West or the more recent wave of "militant tropicality" across the South—was struggling intensely to replace the functionalist certainties it had previously taken comfort from with a form of analysis more in keeping with the real world. When the left did take to the field, shorn of the fraying bell-bottoms of too much theoretical introspection, it was in the guise of a crusading version of the ancient art of humanitarianism. For left-wing activists like Bernard Kouchner, saving lives in the name of social justice was better than feeling oneself obliged to take them in the name of revolution. Others, like Régis Debray, would learn how to get a book or two out of tacking between the revolutionary vanguard and the palace intellectual. In both cases, the politics of it all would prove no less factional.[57]

Meanwhile, the mainstream left, which would in due course bequeath the "third way" projects of Bill Clinton and Tony Blair, was itself busy bedding down in "actually existing liberalism", passively smoking (though not apparently inhaling) the same ideologies of globalisation as were coming from the right: suspicion of the state and anything that smacked of planning and the celebration of wealth as an end in itself. *This* left was firmly committed to fighting tooth and nail for "equality of opportunity"—a state of affairs occurring naturally enough in free-market societies, so hard did they have they to fight for it. But it no longer cared very much for "equality of outcome".

Internationally it was the same story. The new left was now adamant that doctrines were little more than words, and it went to great lengths to strip away the last vestiges of anything that might smack of ideology (unaware it seemed that the new right was running a brisk trade in precisely the opposite). But it was equally inclined to forget that whips are also things, as Michael Ignatieff once put it.[58] That is, until Blair's revanchist Christian internationalism, with its doctrine of the international community, made its entrée at the very end of the century. But whether active or not, the new left for the most part merely completed Robert McNamara's call to defang the Third World, turning it from a wolf to be feared into a street dog to be pitied and spurned.[59]

The authoritarian left was a spent force too. This was apparent long before 1978 and the beginning of Deng Xiaoping's reforms in China. Chinese premier Zhou Enlai had been one of the primary movers behind the more internationalist vision of Bandung. But China was rapidly becoming its own non-aligned movement now.[60] It did not need the Third World any more and, for the time being, no longer had much interest in it. The Soviet Union was also no longer quite the friend to the poor that Khrushchev had claimed, perhaps genuinely, that it always would be. Where was the Soviet Union's voice at that decisive meeting in Cancún in 1981? Answering that question, according to one exasperated delegate, was "like doing a post-mortem without a body".[61]

The socialist camp's greatest appeal to the poor countries had always been that it was free of the taint of colonialism: imperialism was supposed to be the highest stage of capitalism after all. But that appeal and the promises made on the back of it, had all but vanished the moment

Soviet forces crossed the Amu Darya into Afghanistan. Even the rise of Mikhail Gorbachev—a man who saw each nation as in charge of its own destiny, and no fan of the Afghanistan invasion—would change precious little about Soviet foreign policy, from the Third World's point of view. It was no real cost to him, personally, to leave the Third World to the West in order to extract greater concessions for his own domestic empire at home.

THE AGE OF MILLENNIAL DEVELOPMENT

The neoliberal era we inhabit today was forged in the political-economic fires of the 1970s. But it was not until the post–Cold War era that laissez-faire policies became standard operating procedure for a new age of "peace and human rights". Human rights were indeed the last permissible utopia, as the historian Samuel Moyn reminds us. And while they may have burst onto the international scene amidst the tumult of the 1970s, the reason human rights went on to *become* the universally accepted global ethical language they represent today was in no small part because the entire preceding era they burst forth from—an era in which, as we have seen, the rich discovered happiness amidst a world of endemic poverty—had itself done so much to foster the idea of minimum standards as a truly global norm. Human rights today protect and enshrine that view yet more deeply still. It is true that they enable a great many causes—often very noble causes—to be waged in their name. But they do so at a cost to the earlier, welfarist ideals of greater equality of outcome.[62]

This has had profound political implications. The state, particularly after the fall of Eastern European communism, was invariably seen as a barrier to the exercise of those rights, for example. And that meant that about the only institution we had yet devised to ensure both freedom *and* equality found itself in the curious position of being framed as an impediment to human progress, no longer the means of its realisation. Instead, what John Rawls described as a sort of quantitative utilitarian calculus—"the greatest happiness of the greatest number"—embedded itself as the dominant distributional ethic internationally.[63] The claims and experiences of individuals and groups relative to each other now came to matter less than total volume growth. And though they kept themselves largely distinct, human rights and markets nonetheless operated in tandem to ensure this: the one ensuring universal participation the other ensuring the reinforcement of success. In short, by the end of the Cold War, a new era of disembedded liberalism had already been constructed to replace the embedded liberalism of before. For poor countries this meant, most simply, that the "where" of development shifted. The nation was no longer the locus of progress since the state was reportedly "dead". All that was left was the market and those people willing to take part in it.

But if the state was indeed struggling, this was in no small part *because* of the market: a full third of all developing-country governments had turned the corner of 1989 owing more than 200% of their gross domestic product to others.[64] Talk was all about how much aid rich countries should be giving the poor, overlooking almost entirely the fact that far greater sums of wealth were flowing back the other way in the form of debt repayments. The UN report *Adjustment with a Human Face*, published at the end of the 1980s, documented just how unsustainable the situation had become—and this was just before formerly communist Eastern Europe was to join developing countries on the Western bank's ledger sheets.[65] Meanwhile, the World Bank and the IMF kept tut-tutting and recycling their advice from "getting the prices right" (monetarism) to "getting the macropolitics right" (deregulation) to "getting the enabling environment right'" (the adoption of Western norms and standards).[66] The lessons of the global South having been applied to Eastern Europe, the lessons of Eastern Europe—particularly regarding the need to focus on values like "democracy"—were soon being applied right back to the poor world.

Since then, it has been as if, no longer requiring a positive self-image to foist against the evils of communism, we simply got used to not needing one at all. We certainly got used to exploring our opportunities rather than observing our responsibilities. And so perhaps it is not so surprising that the very conditions for freedom that our opportunities rely upon have fallen away in the meantime. It is certainly no wonder that the level of inequality within nations has begun to creep back up, taking with it, it must be said, the likelihood of political and economic volatility between nations.

In contrast to the post–Cold War era, in the age of Truman people at least felt able to speak of the great suffering of those living in poverty as if it were a collective human tragedy: an indictment of society at large, not the fault of the poor themselves. In our present age of Millennial Development, our attention is elsewhere and our empathy spliced, as we have seen in chapter 2, into what is at once an excessively emotional register of "acting now" and an excessively technical discussion of "poverty alleviation". In both respects we are a good deal less willing than before to consider the fact that poverty elsewhere is but part of a wider social question that includes the rich, and their ways of wealth, as well.[67]

The historian Odd Arne Westad once pointed out that the great tragedy of the Cold War was the way that it saw what were originally two fundamentally anti-colonial projects become part of a much older pattern of domination, thus continuing the work of the colonial powers. As he notes, the people of the poor world being only too aware of this, the appeals to them of Americanisation and socialism waned long before 1989, replaced by a local politics often organised along the lines of the very identities and subjectivities—religious and ethnic radicalism—that the Cold War powers had suppressed in order to promote their vision of the future. The West has since made a growth industry out of the security crises of "failed states" and the sustenance they give to terrorists. But in truth, many of these local movements first turned to terrorism precisely because they did not have a state worthy of the name to capture.[68]

Despite the very serious political consequences of all this, today's poverty talk is decidedly apolitical. "Today, more and more people agree that poverty anywhere is poverty everywhere," said James Wolfensohn (then head of the newly pro-market, anti-state World Bank) in 2003. "Our collective demand is for a global system based on equity, human rights and social justice. Our collective quest for a more equal world is also the quest for long-term peace and security," he continued.[69] All very noble, but hollow too—cavernously, wretchedly hollow. In truth, because of the felt need for the Western model of development to look better than the Soviet's five-year plans, because of the celebration of economic growth this led to, and because of the return of conservatives to international positions of authority from the 1950s onwards, the elements of a distinctly utilitarian international order were hoisted into place, and the immunity of wealth from critique was guaranteed.

The preference, rather, was always for working more directly on the poor, providing basic protections when necessary, or simply preaching at them from afar. The great tragedy of the modern era, then, was that the political structures that had developed around a patrimonial Western order were never given the heave-ho in the same way that the notion of equality of outcome was. Instead, the West embraced a politics of trickle-down growth with human rights, as Moyn says, providing the promise of a minimum floor of protection at the bottom while quite forgetting the need to also construct a ceiling on wealth at the top.[70]

This is the fundamental paradox upon which Millennial Development now pitches its tent at the end of half a century of the making and remaking of the Third World. This is the point at which the UN Sustainable Development Goals, which will continue the agenda of the Millennium Development Goals through to 2030, enter upon the world scene. And *this* is what enables the extreme inequality that we see in the world around us to be overlooked in favour of a preening obsession with extreme poverty. For all their benefits, the MDGs have for nearly fifteen years fretted about patching up the worst of the symptoms of poverty, encouraging in us a tendency towards siloed thinking and a fixation on measurable goals of poverty "reduction". In reality, most issues of poverty are inseparably political and bleed into one another: How can you study if you are too ill to go to school? What if you are ill because you have been exploited at work—or because you no longer have a job to go to?

What is needed in place of this type of thinking is a greater acceptance by the wealthy of this world that they already *do* treat the poor world as part of the same society as they are, just not very often as equals. This is not a moral critique. It is much more obviously a historical critique. And we must now either subject wealth to the same "reduction" to which we have subjected poverty to for fifty years or introduce policies that limit its effects on others. When pushed, we would almost all concede that the humanity of others is a property inherent to their existence, deserved by all at birth. But this is not enough: policies are made through institutions and binding agreements, not through some abstract notion of humanity. And that, in turn, requires a little more recognition by those of us who take such things for granted as to our own *ongoing* encroachments upon the lives of others.

4

THE WAY OF WEALTH

The taxes are indeed very heavy; and if those laid on by the government were the only ones we had to pay, we might more easily discharge them, but we have many others, and much more grievous to some of us. We are taxed twice as much by our idleness, three times as much by our pride, and four times as much by our follies.
—Benjamin Franklin, *The Way to Wealth*

The world of homestead and harbour that Benjamin Franklin wrote for in the eighteenth century is a long way from today's globalised world. But so too is the immediate post–World War II era, when, as we have seen, for a short period a more benign form of internationalism reigned. Since the end of the Cold War, the world has changed dramatically yet again. So much regulation has been abandoned that it has become hard to imagine how some of it could ever be put back into place; demographic changes and urbanisation have together fundamentally reshaped the world; and new global powers like China and Brazil have emerged with their own burgeoning internationalisms, their own versions of the good life, and their own bottom lines.

But amidst all this change, much remains the same. How much more concerned, really, is China today than Europe ever was about putting Africa's interests before its own? To what extent will Brazilian elites be committed to their own poor as that country's growth rate inevitably slows? Different nations may now have an influence on global trade rules (the G20 and not just the G8), but the underlying norms that guide those rules remain the same. At the end of the day, neither the "imperatives" of globalisation nor the much-regretted "elusiveness" of development will ever present us with anything more than choices. Our problems, and our potential, stem from what we ourselves then make of them.

TRADING PLACES

Those choices, we have seen, have a history. But they also have a geography. As the "debate" over aid and development illustrates, we have learned to do as much good for others as avoids jeopardising any of our own special privileges. There are many who would criticise the morality of this, and more who would defend it. But why not just challenge the logic of the trade-off itself? Why are we so sure that it is by *not* doing more for others that we preserve our own good fortune? We cannot alter the fact that self-interest guides our choices, and we may not even want to, but we might enlighten the terms of that self-interest a little.

It is anyway in our interest to do so. It was the same accounting techniques developed to supply the poor world with risky credit in the 1980s that were later used to offer excessively leveraged mortgages to the under-waged of America in the 1990s, ultimately precipitating the credit crisis. Similarly, the manner in which the global North chose to rid itself of a problem of inflation at the end of the 1970s (by converting it into a crisis of debt for the global South) is mirrored today in the way that the national debt taken on by the British government to save its banks (and through them the global financial industry) is gradually—through the push of austerity and the pull of access-to-credit schemes—converted into private household debt owned by people themselves.

In light of this, and for all that our politicians have been desperately seeking to reassure us that things are getting back to normal since the financial crisis of 2007 and 2008, it is far from clear that normal is actually where we want to be. The German pharmaceutical company Bayer, for example, recently sought to revoke a compulsory license that had been issued on public health grounds by the Government of India for one of the company's anti-cancer drugs, Sorafenib. Novartis, a Swiss pharmaceutical giant, then went further and took India's government to court over its domestic intellectual property law, a law Novartis disliked because it prevented it from selling its own anti-cancer drugs at monopoly prices. Both companies did these things because they assumed that there exists an informal pecking order in international affairs, whereby Western corporate interests outrank those of non-Western civilians. Fortunately, they each lost their respective battles in the courts this time around. But they were nonetheless right about the wider state of affairs.

It is not just the pharmaceutical industry, of course. In Britain Nestlé is one of a leading group of companies to have committed to paying all of its employees a living wage (which is greater than the legally acceptable minimum wage): an admirable step. But in Brazil Nestlé employs a boat to chug up and down the Amazon, hawking packaged food and ice cream to remote communities along its banks. Elsewhere it supplies local micro-entrepreneurs with its products and has them administer IOUs to folks who cannot afford them. In India, Unilever's door-to-door salespeople promote its products like underpaid Avon ladies, while in Africa SABMiller is pushing cheap, "entry model" beer to the world's thirsty. It is quite wrong, then, to say that the global poor are being forgotten or overlooked. To the contrary, they are capitalism's new frontier, where Western corporations hope to sustain their shareholder value even as they present a responsible corporate face back home. And what is happening on that frontier is something it is worth paying attention to.

Of the many possible responses to Apple's poor treatment of workers at its Foxconn factory in China, surprise was arguably not one of them. The media were right to make a cause célèbre of the issue, but had the media really done their homework, Apple might equally have been called up for its failure to adequately ensure that its sourcing of coltan from the conflict-stricken Democratic Republic of Congo is not aggravating local violence.[1] Its trade there could certainly benefit from something like the Kimberley Process Certification Scheme, which seeks to regulate the diamond trade (especially given the links between the Rwandan Patriotic Front and Rwanda Metals, which supplies Apple's suppliers).

But it is often the more quietly pervasive forms of injustice that do the greatest damage. It took the collapse of a ten-storey building in Savar, Bangladesh, where clothes for some of the largest Western retail outlets, like Primark, and some of the most popular brands, like H&M, were made, for these sorts of practices to be pushed to the forefront of the wealthy world's consciousness. When the building at Rana Plaza gave out, taking more than a thousand lives with it (imagine the reporting if that had been a terrorist action), the commodity chain linking purchaser to producer also temporarily collapsed. People in the rich world were for once forced to see the *real* cost of the clothes they are not prepared to pay market value for. The extent to which such a tragedy could have been avoided became clear.

We perhaps take too literally the claim that the only logic behind the market is the invisible hand of countless individual decisions made simultaneously—as if there were no politics there in the first place, guiding what those hands can do. Indeed, when we think of something like "state capitalism", it conjures for us an image of authoritarianism that we immediately project onto other places: China, Russia, Venezuela perhaps. It is as if some of us do not see the state at work in markets everywhere, from the granting of Nigerian oil concessions to British oil companies before independence in 1960, to the great carve-up of Iraqi oil interests today, where the American state is ever present, for all the Iraqi one is failing, opening doors, making introductions, pressing contracts.[2]

When we take the trouble to look, in fact, it soon becomes clear that for all the free-market rhetoric that states are best kept in the background (the better to unleash our creative, entrepreneurial spirits), the rich world's corporations and financial markets have long *required* states to accord them special favours in order that they can function. This is how corporations gain leverage over poorer countries so as to provide us with the value we demand as Western consumers. "Virgin would have been just half the company it is today", says Richard Branson in defence of the blind eye the British government has long turned to the company's record of persistent tax avoidance. He is right, of course, but so too would British public services have been that much better funded over the past twenty years. The same applies to the UK High Street retailer Boots, which withheld up to £1.2 billion in taxes over recent years—enough to fund the entire nation's dispensary requirements for two and a half years—despite having a special deal with the National Health Service to provide medicines over the counter.

Such practices, then, are not unique to the poor world, but they do disproportionately affect the poor, wherever they may be. What happens in some industries also simply matters more for society at large because of the nature of those industries. The global arms trade is a case in point. Three-quarters of the annual $50 billion in global arms sales come from rich countries selling to poor ones, with 85% of the proceeds going to the five permanent members of the UN Security Council. And yet there is nothing like the anti-proliferation or nuclear disarmament movement of the 1980s confronting such destructive trade today. Perhaps this is because the weapons are invariably used elsewhere. The past ten years, however, have been an object lesson in why we should not consider ourselves entirely immune to the consequences of this trade.

Dealing in arms is pertinent in another sense too. Manufacturing weapons has always been the West's preferred way to claw itself out of economic trouble: it was, after all, the rise of the military-industrial complex that helped put the US economy back on its feet after the Great Depression, and it did so again in the 1970s. This remains the case today, as British prime minister David Cameron was all too quick to acknowledge by undertaking a whistle-stop tour of the key trading nations of the UK arms industry as part of his government's wider economic recovery strategy.[3] It is as if arms sales didn't rank among the chief causes of corruption in the poor world. Or as if, before even a shot is fired, the money used to buy weapons weren't such a sizeable chunk of poor countries' welfare, health, and education budgets. And shots do get fired, which is why arms sales are also a major cause of civilian deaths in poor countries. As Dwight Eisenhower said, "The world in arms is not spending money alone. It is spending the sweat of its labourers, the genius of its scientists, the hopes of its children."[4] But, of course, what we should say is that the world in arms is spending the sweat of *some* of its labourers and the hopes of *some* of its children.

Usually the line taken here is, "Not our problem." And "not our problem" was also the West's official response to the Asian financial crisis of 1997 and 1998: an event that took the sheen off the "emerging markets" story that had captivated financial interests in the immediate post–Cold War years. There is plenty of evidence emerging that the capital inflows of financial market investment in poor countries actually reduce substantive local investment opportunities because of the distortions they can cause in the exchange rate, among other reasons.[5] So it was perhaps not by coincidence, after all, that those countries and regions most favoured by emerging-market status prior to the crisis—Mexico, Russia, East Asia, Argentina—all were the first to be plunged right back into debt, giving the International Monetary Fund grounds to step in with "conditional" loans once more, at which point the whole sorry cycle was born anew.[6]

The West's response to this was for the most part to blame the countries themselves: it was a problem of "crony capitalism", said Larry Summers, US Treasury undersecretary for international affairs. "The lesson of the Asian crisis is that it is better to invest in countries where you have openness, independent central banks, properly functioning financial systems and independent courts, where you do not have to bribe or rely on favours from those in power," opined Tony Blair.[7] One is tempted to reply "LNM Holdings!"[8] But in any case the inherent risks of over-leveraging by Western banks were more directly the cause. Had *that* lesson been learned, the next financial crisis—one that struck closer to home in 2007–2008—might have been avoided.

What all of this points to is that one of the principal problems confronting poor countries today is the distinct lack of oversight by those who could, or should, be regulating global flows of every sort. The existence of tax havens is amongst the clearest examples of this. No fewer than ninety-eight of the companies listed on London's FTSE 100 Index in 2011 used tax havens as part of their standard business model.[9] So it should not have taken a few public relations blunders from High Street companies like Starbucks to wake us up to this.

It was well known, and long ago, that Democratic Republic of Congo's Mobutu was sending his ill-gotten gains to Switzerland. But it is equally well publicised today by organisations such as ActionAid that Western corporations like Glencore (which has avoided paying $76 million in taxes to Zambia) and, again, SABMiller (which shifted $100 million in profits off its taxable accounts in Africa) are no less systematically depriving the poor world's public of their due by avoiding paying taxes.[10] Trade mispricing—altering import and export values to reduce tax liabilities—costs African nations alone around $38 billion a year, roughly equivalent to the United Kingdom's annual overseas development aid budget.[11]

Tax dodging affects all countries, of course, but it too affects poor countries most of all: to the tune of $160 billion globally every year, according to Christian Aid. Like corruption, tax havens are nothing new: between 1970 and 2008, for example, they ensured that up to sixty cents of every dollar of foreign loans made to poor countries went straight out of those countries.[12] But they have gotten worse over time. Tax policy matters, including the taxes you can levy on a foreign company. If you can't collect taxes from your rich or if your poor work in informal sectors, then your only hope of an income with which to run basic services is to be able to tax those corporations that make profits in your territory. According to one estimate, addressing international tax loopholes would be sufficient to pay for a reduction in younger-than-five mortality of one thousand children per day: all without a cent more in aid and no change in existing political systems.

Yet what chance is there really that such policies will be introduced when the world's largest corporations pay the major accountancy firms handsomely to continue smudging out the already-diaphanous line between tax evasion and tax avoidance? Or when governments themselves merely reiterate to their electorates their "firm belief" that there is a difference between the two (while studiously acting otherwise)? Perhaps this is to be expected: the British economy survives in no small part because London is a glorified tax haven. But when the British courts support the government's claim that its tax deal with Starbucks (a company which on paper claims a *loss* in every single one of its UK High Street branches), is a rigorous one, then one begins to understand where that same government gets the gall to claim that it needs the people of Britain personally to shoulder deeper cuts so that the government can "balance" its own ill-kept books.

Tax dodging has touched a nerve in Western countries of late. But for poor countries it is just one example of a much wider problem of unregulated capital flight. Transfer price fixing, in the appropriate jargon, is the corporate practice of juggling firms' costs and benefits so as to report profits in those corporate entities in the lowest-tax destinations, where a company might have an office, or a postbox, but no real connection, and so avoid paying it where they do. There is nothing illegal about this. But given that 68% of all global trade occurs *within* multinational companies, it is clear just how little world trade is actually regulated. When today's market liberals talk of the need to push away the dead hand of regulation, then, it is increasingly unclear where this regulation that they are apparently so upset about is actually to be found.

The real point, though, is this: all of these irregularities and inconsistencies are well documented, and they could all change given the right policies and the necessary political will. But market optimists—what else are we to call those who call for freer markets after the credit crisis—still like to believe, as Doctor Pangloss, that all is for the best in the best of all possible worlds. We are all of us, however, guilty of letting such arguments go unchallenged just because, as we say to ourselves ad infinitum, what can any of us on our own actually *do* about any of this?

FAILING STATES

The whole point of having states is precisely so as to be able to address problems like this. Yet as the Princeton economist Dani Rodrik puts it, "Talk about re-empowering the nation-state [today] and respectable people run for cover, as if one has proposed reviving the plague."[13] Not all states are the same, of course. Politicians are usually quick to point out that poor states are not as good as our own rich states: they are less democratic, less efficient, less willing or able to protect their citizens. But this is to remain within the "territorial trap" of assuming that state sovereignty lines up neatly with state borders: that other citizens need protecting only from their own states, for example, or that part of the reason those states are incompetent in the first place is the influence upon them of more powerful ones.[14]

For all that the fear of foreigners is back with a vengeance in the West, brought to us by such purveyors of cosmopolitan goodwill as the UK Independence Party's Nigel Farage or the Front National's Marine Le Pen in France—politicians who fear everything and anything that wasn't made this side of the beach—it is usually those foreign nations that have greater cause to fear what comes from us: be it drones, trade embargoes, the destructive power of bond markets, or even diseases, as with the cholera introduced to post-earthquake Haiti. Or maybe just the way that our management of things like diseases slips into the management of the peoples and nations those things are presumed to come from, because doing so seems easier. The recent global response to Ebola is a case in point: seal up the borders, stop it in its tracks. There have even been suggestions that drones might be used in the Ebola response—the better to limit the risk to ourselves, albeit not a very touching form of health care.[15]

It is just this sort of mind-set that lies behind the current international obsession over so-called failed states. To talk of a failed state presupposes, of course, that a state existed before and that people want it to exist again. In the case of Afghanistan, perhaps the failed state par excellence, the former may actually—against the odds of geography and history—be the case, but the latter is far from clear. Afghanistan is thus asked to be a liberal democratic-looking state today—and it fails in relation to that demand—only before the international community. Domestically, there is nothing to suggest that the regime in Kabul, which scarcely collects any taxes, or even has much domestic legitimacy, is much wanted at all.

The difficult but important questions this raises are swept under the carpet, however, because what matters most to the Western leaders who still dominate the international agenda is not whether Afghanistan, outside of media-saturated Kabul, is well governed, but what the instability of Afghanistan means for *us* (a concern that boils down to terrorists and drugs for the most part). Hence, the UN Office on Drugs and Crime acts on behalf of Concerned of Tunbridge Wellses the world over to strongly encourage (very strongly, as it happens) Afghan farmers to desist in growing opium poppies—Afghanistan being one of the world's primary suppliers of opium. It has been shown time and again, however, that farmers will return to poppies, whose cultivation they learned from their parents, and which they can rely upon in the difficult economic circumstances that we in the West have dropped them in. And with Afghanistan having for decades been more properly governed at the level of individual valleys, the policies concocted in

Kabul anyway have even less meaning for most Afghan farmers than do the policies emanating from Brussels have for people who vote for the UK Independence Party.

It is not just the domestic politics of so-called failed states that our leaders and opinion setters have a tendency to overlook today. They rarely think very much either about the historical and geographical points of connection between our own societies and those we are busy rediscovering in the light of our present obsessions. In the lead-up to Iraq, for example, Thomas Barnett popularised, first in the pages of *Esquire* magazine and then in a *New York Times* best-selling book, the virtues of the Pentagon's then "new map of the world".[16] Barnett took this as the foundation of a new geography of "war and peace in the twenty-first century" (the subtitle of his book) and wrote approvingly of the way this imaginative cartography found its inspiration in the writings of Victorian imperialist Halford Mackinder. The map presented a world divided into a "functioning core" and a "non-integrating gap": the latter a sort of global-scale gerrymander identified on the tautological basis of those areas where American forces have been involved in the post–Cold War period.

In such imagery, and in the writings of journalists like Robert Kaplan, Paul Bracken, and Thomas Friedman, who unthinkingly deploy these ideas, the world is reconfigured into "shatter zones" and "belts" of human misery. Great swathes of generalisation, that is to say. Such maps hoodwink people into believing that there are parts of the world that are somehow inherently "off-limits" or "beyond the reach of law": places like the borderlands of Afghanistan and Pakistan (which become, in pseudo-military jargon, the AfPak borderlands), or even, for that matter, Guantánamo Bay, where the needle of military interrogation was for some time said to be permissible and Western legal formalities an encumbrance: until the US Supreme Court pointed out that since America had run the place for a hundred years, its own rules—and not those of nineteenth-century Cuba, from which it borrowed the bay in perpetuity—in fact applied there. (The Bush administration was, needless to say, aghast.)

One of the consequences of failed states is that people do indeed choose to pack up what bags they have and to leave for more promising lands—though in remarkably fewer numbers than one might expect. For the receiving countries, this raises the problem of asylum and migration. In these areas in particular the modern Western state reveals its not-inconsiderable capacity to influence the quality of lives lived elsewhere through the policies it implements at home. It is claimed today that class matters less than geography, for example. This may be true in the sense that *where* you are born, statistically speaking, is likely to condition your life chances and opportunities (as World Bank economist Branko Milanovic has argued, the location of your birth and the social status of your parents together account for 80% of the variability in global income).[17] But we miss a large part of the picture if we overlook the extent to which that geography is itself a reflection of prior uneven relationships of power. At the end of the day geography is enlisted in the *service* of class.

There is no doubt, however, that migration from the poor world to the wealthy (or wealthier) world is one of the central challenges of our time: one that will increase before it goes away. More's the pity, then, that it is being met for the most part by thick-skulled defensiveness. In the United States the conservative House Immigration Reform Caucus has dug deeply into its intellectual pockets and pulled out the idea of having troops be stationed on the border with Mexico to address the problem. Europe has opted instead for an archipelago of refugee camps and asylum processing centres within the Schengen Area and a greater level of surveillance without.

On both sides of the Atlantic, however, our confusion is such that we don't even know what to call a border any more, and sometimes we aren't even sure where it lies: the border between Israel and Palestine is a "defence wall" that repeatedly breaches the Palestinian side of the border; the increasingly blurred line that runs between the United States and Mexico is a "trans-cultural borderland" marked by drone flyovers and heavy-handed policing; and Europe's Mediterranean Sea border, patrolled by Frontex-operated ships, has for several years been creeping on to the beaches of northern Africa, as countries like Libya and Morocco are given migration management privileges on Europe's behalf: the better to assume migrants guilty until they can prove themselves innocent.[18]

We are almost entirely without perspective here. For all that borders are one of the things that national governments are understandably most sensitive about, and for all that borders are certainly made to look and feel that way for those who daily risk their lives trying to get across them and into the rich world, it is only since the end of the First World War that one even needed a passport to travel. Surely our memories are not so short that we cannot devise at least a little creativity here (though hopefully not of the sort used to turn Heathrow Airport's new Terminal 5 into a luxury shopping mall for some and a privately run immigration detention centre for others—where an eighty-four-year-old man died last year with his hands still cuffed behind his back).[19]

We could all be a little more attentive to the fact that, from the perspective of the poor world, stricter migration laws for those denied the privilege of snaking their way past endless shelves of Paco Rabanne, who are instead made to sit in an airless detention room while not just their papers but also their life stories are checked, act as a fundamental infringement of the right to freedom of movement and a brake on the free-market protocols they have been preached to about by the very countries in whose secure "holding rooms" they are made to wait.

THE INTERNATIONAL "COMMUNITY"

For decades international relations scholars have insisted that the world beyond the nation-state is an anarchic swamp, a political no-man's-land. Yet compare the travels of a migrant as he negotiates a path out of Afghanistan to Iran and then travels onwards to Dubai where he works for around $100 a month and sleeps rough at night, to a business traveller porting a Nexus "smart entry" card at the US-Canada border at the start of the day's commute, and it soon becomes clear that, far from being a political no-man's-land, the space of the world is in fact daily bent to the whims and the will of the powerful. Squat as a fly on the wall in the meetings to decide who would replace the disgraced Dominique Strauss-Kahn as managing director of the IMF in 2011, and the meaninglessness of the notion of the swamp will likewise become clear. The problem is not that the international realm is ungovernable, but that the spaces and institutions that constitute it are subject to different forms of government, and many of those are at present irresponsible or ineffective, or both.

The brouhaha over Strauss-Kahn's replacement was a reminder, above all, of a characteristic irony of the times in which we live: that while we are encouraged to debate openly and in public at the margins of a given problem, the things that really matter are decided upon in private, if not behind a wall of tear-gas-armed security guards. As against the public debacle that was the Climate Change Conference at Copenhagen, negotiations between the twelve would-be member nations of the Trans-Pacific Partnership have been carried out comparatively smoothly and with precious little media scrutiny to date. This despite the fact that, as concerns US citizens, for example, the TPP "would allow foreign companies to sue the United States government for actions that undermine their investment 'expectations'". With parts of the trade deal not due to be declassified until four years after the TPP comes into force, it is little wonder some groups refer to it as "NAFTA on steroids".[20] And yet what is most worrying here is that the TPP is not an isolated disgrace to democracy. It is just one more agreement in a long and growing line of regional treaties and accords which are mopping up the middle ground between states and the international system and claiming this ground in the name of the market. The Trans-Atlantic Trade and Investment Partnership, a US-EU effort to resurrect the spirit of the previously defeated Multilateral Agreement on Investment, is another.

Regional economic gerrymanders of this sort have been in vogue ever since the end of the Cold War. But their current rollout around the globe (as with the various free-trade agreements that the European Union is seeking to enter into with India and a cluster of Latin American states) is worrying. Along with tariff regimes protecting American steel and European agriculture, such agreements ensure that the rules that govern the international system today are the very opposite of the benign international environment of the earlier postwar era. They are no longer even a means of locking in the interests of rich country citizens as against those of poorer countries. Today they are a means of locking in corporate interests from many nations over those of citizens everywhere.

Hope may well spring eternal, but opportunities for sorting out this sort of pre-emptive skewing of the playing field do not. Already today there are literally thousands of bilateral and regional trade agreements that have been put into place around the world in addition to the more usually criticised World Trade Organisation framework itself. Of course, if the WTO did what it was supposed to do, we should all be grateful indeed. But the WTO we have is one whose mechanisms have been rigged in advance via such "agreements" as the General Agreement on Tariffs and Trade, in which poor nations must commit to reduced tariff and capital exchange controls as a basic condition of trade. This is a requirement that would have ensured that America never grew beyond Virginia in its time. The WTO would survive its excision.

And yet we cannot throw up our arms in consternation at corporate malpractice and the wider "free trade" agenda here. There is no point. Corporations are only pushing at the limits of what they are allowed to do for the most part. The real problem is that for all we live in an economically globalised world, we do not live in anything like a sufficiently politically globalised world. It is for this reason, and no other, that the international political system today suffers from a democratic deficit. And until we address this, until the public gives itself a voice at the negotiating table of planned trade deals like the TPP, we will continue to see our autonomy as national citizens dug out from underneath us.

We will also continue to see a world in which global politics more broadly respects only the most powerful among us. Even the United Nations is not immune, with the five permanent members of the Security Council acting more like a global police officer than anything else. There are those who say that the rise of NGOs offers the solution to this: that we don't need some parastatal body like the United Nations getting under our noses, anyway. This is the voice of the micro-entrepreneur speaking. Certainly it is true that NGOs, and not states or international bodies, have become the "first responders" of public morality in the global age.[21] But is this a good thing? For their proliferation is an expression too of our wider belief that the question of which issues to support has become, like taking part in the proverbial charity run, a matter of personal choice. It isn't clear why we have come to think this. But it is clear that as a result we have become even more reluctant to consider any problems we aren't personally "invested" in ourselves.

All this is inevitable in the absence of a sufficiently matured democratic international system. And it leads to a lot of chasing of one's own tail, as the systems of oversight and delimitation necessary to effective government must perforce be ushered in via the back door. To wit, today's endless discussions on "governance"—a term that the left most frequently indulges in, seemingly having forgotten that it first emerged from the new right's critique of the interventionist welfare state. When words like "good governance", "accountability", and "democratisation" are thrown about quite as often as today they are, one is tempted to suspect that we doth protest too much indeed.

During the past forty years, for example, the World Bank first turned its attention away from short-term project financing to longer-term programme financing, and eventually to a whole range of social and political prescriptions addressing everything from budgetary discipline to human rights and the appropriate business environment.[22] It has moved in short from helping countries to get a leg up to telling them what to do. As former vice president of the World Bank Robert Garner put it to one of the bank's young firebrands in the 1950s, "Damn it, Lauch! We can't go messing around with education and health. We're a *bank*!"[23] That is not a message we'll hear from today's World Bank vice president. It might be worthwhile asking him why.

BE VERY AFRAID

We are not well served by our leaders in any of this. Yet there is one thing that *will* get our current crop of politicians—be they presidents or prime ministers, board chairs or secretaries-general—all singing together: the catchall term "security". Where previous generations lived with constant (and useful) chatter about the need for full employment versus the threat of inflation or were encouraged to engage in the War on Poverty, our generation—the generation of the "risk society"—is encouraged to preoccupy itself with being secure, something that is rarely an incentive to generosity and is frequently an existential dead-end to boot.

The US Republican Tom Ridge—presumably choosing his words carefully on being charged with running "homeland security" in the months when America slipped from being a nation to a homeland in official public discourse—said that the new grand task of the nation was the protection of "essential liberty".[24] This was a task for which all Americans were now responsible. And to achieve it, security would become the new national mission, likened, he said, to the Transcontinental Railroad and putting a man on the moon. Really? In any case, this was not the sort of security envisaged by Franklin Roosevelt in the 1930s, at the beginning of a period when national interest and the general welfare of society merged.[25] It was not what Roosevelt envisaged in 1944 either, when in speaking of a "family of Nations" he sought to extend the vision of the New Deal—the belief that "freedom from fear is eternally linked with freedom from want"—to the rest of the world.[26]

Today's understanding of security is quite different to this. Sustained by the tireless efforts of what the French sociologist Didier Bigo calls "the managers of unease", it has become the wealthy world's "master narrative", directing citizens this way and that, telling them to mind not just their own business but other people's too—all in the name of policies which are themselves, as one insightful critic puts it, "predisposed towards the exercise of violence in defense of the established order".[27]

Security policy has thus become exactly the constellation of force arraigned on behalf of privilege that was predicted long ago, and not by Karl Marx, but by Adam Smith: "Civil government," Smith wrote in *The Wealth of Nations*, "so far as it is instituted for the security of property, is in reality instituted for the defense of the rich against the poor, or of those who have some property against those who have none at all."[28] Security, make no mistake, is about class. And in its post–Cold War guise consumes the money we might otherwise spend on things like preventive health care and channels it instead to the interests of private military contractors and the purveyors of zero-tolerance policing. We are told to be afraid, in short, because it makes some people richer when we are.

But it also undermines our democracies. While many aspects of the state have been in retreat over the past thirty-five years, for example, the corrections industry has consistently been the fastest-growing sector across all three levels of the US political system (county, state, and federal). White-collar crime today costs the United States several hundred billion dollars each year, with scandals like those that rocked Enron in 2001, WorldCom in 2002, and Bernie Madoff in 2008. But it is overwhelmingly blacks, Latinos, and unemployed whites who have been sucked into the country's growing prison-industrial complex.

A more punitive form of government in the name of security is also increasingly deployed internationally (in the age of Roosevelt the preference was for welfare). Before all of our eyes the global war on terror has become a "planetary security effort", in the words of one critic: a source of constant intrusion, disruption, and ultimately violence for those labouring to get on with their lives in countries that come on to the Pentagon's or MI6's radar.[29] At the same time, reducing *their* problems to prior questions of *our* security has become an ever more powerful means for the rich world to distance itself from the realities of poverty and its discontents. To make the world more just we have to act with greater justice, which means we have to open ourselves up to others. This is the very opposite of security.

The West is increasingly losing whatever moral grounds it might once have had to question the nature and causes of conflicts elsewhere in the world: be it China's bullying in Nepal or the current ethnic imbroglio that is Syria. Knowing this, is it any surprise that some countries pursue domestic military conflicts today with greater disregard to international opinion—with the literal expectation of getting away with murder? When Israel targeted 450 sites and individuals in Gaza in mid-November 2012, for example, of the first five BBC headlines on the event, two presented the attacks as "Israel's Gaza rocket problem". When BBC Radio 4 then reported that Israel would suspend "military operations" for the Egyptian prime minister's visit, it did not once use the word "violence" or "attacks", but merely "operations".

Amidst all this double-speak it is no surprise that Western leaders struggle to cling to the moral high ground. Yet they are reluctant to abandon the position altogether, since it serves them far too well in times of crisis—times we are hardly short of these days. NATO acted with increasing entrepreneurial verve under its former secretary-general Anders Fogh Rasmussen to loosen the definition of violence (the better to allow greater room for manoeuvre when deciding whether and how to intervene in countries like Libya). These are moves met in the opposite direction by humanitarian organisations seeking to expand their definition of violence, if not their market share of the world's suffering with it, and the effect has been to undermine the clarity of international law at just the historical moment when it has been needed most of all.

The associated blurring of the boundary between humanitarian and military interventions has, inevitably enough, carried Western nations into the fraught waters of the responsibility-to-protect doctrine. Enshrined as permissible international practice by NATO in Kosovo and under UN Security Council authorisation in East Timor in the late 1990s, interventions by multilateral coalitions of the willing are now "just"—in the words of Tony Blair, and he should know—to the extent that they are based not on military or political objectives in the first instance but on "values".

Little wonder, then, that by the mid-2000s development agencies and humanitarian organisations were crying foul that the military was simply using them as "force multipliers". They were right, of course. But to some extent they had only themselves to blame: for they had themselves plunged with eager hands into peace-building and protection for people who apparently could not be trusted to govern themselves. Imperialism now rides the coattails of the humanitarians; it once was the other way around—yet for those on the receiving end, it is far from clear that the change in roles makes all that much difference at all.

What is all too often overlooked today, amidst all the hand-wringing over our duty to intervene in crises elsewhere—to do *something*—is any consideration of the fact that we might bear some of the responsibility for what has brought events to such a sorry state in the first place. In the way we then *do* go about "protecting" the world's poor—with the current mantra of "protection of civilians", for example—we frequently open things up for more, not less, violence. There is more than a little irony in this. For as my colleague Ole Jacob Sending has observed, if it is true that in the current vogue for "ethical consumerism" the obligation to care and the desire to consume are collapsed into each other (help the poor by buying a cup of coffee or by purchasing a Louis Vuitton bag), the same is no less true in the political domain, as internationally, our promises of "protection" are collapsed into a simultaneous effort to order the world's societies, and its failed states especially, in a manner that is to our liking.

It is here that the logic of security proves its greatest worth for wealthy and powerful states, precisely because it encourages us not to notice connections such as this. But the logic of security also lies behind the revived fortunes of development. For much of the 1990s development had languished somewhat in the absence of superpower sponsorship. Official development assistance dropped off substantially: US contributions halved, in fact, from $16.2 billion to $8.4 billion between 1990 and 1997.[30] Its fortunes revived after 9/11, however, albeit by responding to a rather different tune. Development today is less about human flourishing than it once was, and more about the maintenance of a fragile—and culturally specific—vision of global order, for all that a great many people work in good faith to achieve rather more than this.

Development, as an adjunct to security—a holy grail that goes by the name of "human security"—is one way in which well-intentioned people have sought to square this circle in recent years. But the concept has itself now become a means to reshape poorer societies in light of the rich world's own particular needs and so in many ways it has merely reinforced the problem. High up on the list of those needs, of course, is the desire to calm our anxieties about the flux and flows of an interdependent world. Development and humanitarian actors must now accordingly address themselves not to people in need but to "situations of crisis" and "complex emergencies", and of course to the donors who are paying for all of this. The recent restructuring of the UK aid budget to focus primarily on only those countries that are deemed to be security risks to the United Kingdom says it all, really.

The old Cold War policy of containment—fencing in the Soviets and waiting them out in the hope that they would croak first—has in today's world been transformed into one of selective engagement. To some extent, this is inevitable and sensible. The footprint of our own societies is too large to avoid engaging with the places where the rare-earth metals we require in our laptops are mined and where our food is increasingly grown. But the terms on which we do so are so myopic that we frequently create more of the problems that we purport to solve.

Such realities ought to raise doubts as to the counsel, in this respect at least, of Michael Ignatieff, who in channelling none less than King Lear, proposes that we "'dig up the heath', put [it] under the sovereignty of a nation armed and capable of protecting its own people". This is mixed advice indeed. For we are already digging up the heath, and for our own gain. We are already putting it under the sovereignty of the international community and its coalition of leading states. And we are doing so in accordance with an increasingly corporate jargon that has so permeated national and international political discourse that we can today find it within ourselves to speak of government only as a matter of "efficiency" and "choice", as if the prior questions—efficiency and choice of "what" exactly—were nothing worth troubling ourselves about. It isn't clear that any of this is protecting the people on the heath at all.

THE GROUND BENEATH THEIR FEET

The law is of course a special sort of institution, and developing fair and effective laws should be a priority of any world that aspires not to be mired in inequality. Yet the law is presently failing the world's poor by affording them far too little protection when it comes to matters of trade policy or the economic relations between nations. Where the law is considered to have a central, public role to play today is in the context of human rights, which activists raise as a defence to exploitation in the hope that either marginalised people will claim the rights they are offered, or others will find the goodwill to do this for them. But individual rights without wider social and economic protections are little more than "paper parapets", as James Madison famously put it. Moreover, the law is as easily exploited as enforced by the powerful, perhaps because too many of our leaders trained as lawyers and they therefore feel they know better.

The failings of international law today are a reminder that it is not just the heath, as Ignatieff would have it, but the rule book of precedent and procedure that has been torn up of late. As Britain's Lord Rooker, then UK Home Office minister, put it in November 2001: "The terrorists rewrote the rule book, and we have to do the same."[31] And for the past decade that is precisely what we have done: a process that reached its present high-water mark with the spectacle of a sitting American president outlining a doctrine of war in his Nobel Peace Prize acceptance speech. Such rhetoric as Obama saw fit to deploy in Oslo—"the full force of our justice"—belies the bellicosity that *inflames* global tensions. It is not just the law but also the spirit of the law that matters. In this case, Obama the lawyer really should have known better.

But the law is demeaned in more mundane ways too, such as by the operations of the Orwellian-titled distressed debt investors, more popularly, and accurately, known as vulture funds. Vulture funds, run from regent offices in Kensington and Mayfair, tap into poor countries' wealth by exploiting rich countries' laws and courts. The "accepted" practice here is for funds to buy up bad debt in poor countries, which they sit on while the debt rises. Later, when market conditions are favourable, they take the indebted country to court for the full sum, plus fees. A British court, for example, awarded the American investor Michael "Goldfinger" Sheehan $20 billion in repayment costs against Zambia in 2009. In light of the attitude that British courts seemingly take to what is by any definition a demand for ransom, one is reminded of just such *conscience juridique* as was displayed by the international lawyers flocking about the Berlin Conference of 1883–1884, pardoning in advance their masters' thirst for territory.[32]

There remains, unfortunately, substantial opposition to the idea of an international law that is anything more than a rich man's club, and not just in the panelled offices of Mayfair. Even before 9/11, John Bolton, who would later become US undersecretary of state, said that international legal agreements and binding contractual duties—the Kyoto Protocol, for example—were nothing less than a "worldwide cartelization of governments and interest groups".[33] He could not have better channelled the spirit of Moynihan if he had tried.

Where the real cartels are at work, however, is in the almost invisible fabric of international standards and norms that set the terms of engagement in the global economy. Promoted via everything from ISO standards to trademark practices and intellectual property law, these silent normalisers do an essential job in enabling international harmonisation. But they also tend to hardwire into the international system a set of norms and rules that favours the rich. They create a world in which poorer countries that want to have a stake in the global marketplace must conform to the practices established by those who got there first, whether or not those practices actually work for them.

When it comes to things like producing generic and affordable medicines for consumption by the poor, the costs of a Western-dominated international intellectual property rights regime is measured in actual human lives. But again, the solutions we require here are political and not economic in the first instance. The International Labour Organisation, for example, does a hugely important job in promoting standards of "decent work" around the world. But compliance with its conventions is, at the end of the day, down to national political debate and international pressure.

"A RIGHT DISPOSITION"

Despite what we might hear on the news, we in fact live in a more peaceful world than ever before. Armed conflict within countries—by far the greatest cause of deaths globally—increased steadily over the past fifty years, peaking in 1992. Ever since then, the figure has more or less been on the decline.[34] But what does this formal status of peace really mean in light of the fact that, "universal and lasting peace can be established only if it is based upon social justice", as the preamble to the ILOs constitution reminds us?[35] In the years since 1989, democracy and human rights have come to define what has been referred to as the post–Cold War peace. But for all that America and Europe like to harp on about democracy, it is becoming increasingly less clear what they actually mean by this.

What is clear is that, like the lawyers, they do not consider social and economic rights to be anything like as important as they once were. Mozambique's first Action Plan to Reduce Absolute Poverty (2001–2005), to take just one country, was almost entirely in hock to foreign donor strategies, and foreign consultants largely decided on its replacement (2006–2010): donors were again invited to comment, but not the Mozambican parliament.[36] Really this was nothing unusual. The Millennium Development Goals and the Sustainable Development Goals have been organised in a not-dissimilar way. The underlying problem here is that democracy is rarely understood to mean democratic decision-making over the way that our economies, be they national or global, are run—even though the prioritisation of social and economic rights is supposed to be a core part of the UN's mandate. Instead, the West prefers to deploy the vernacular of human rights as a way to constrain the power of (other) states: though as at least one study has shown, across the twentieth century and all political systems, it is democracies that are most likely to target civilians, and it is in the name of other peoples' human rights, of course, that we today launch humanitarian interventions and the no-fly zones required to support them.[37]

Yet none of this is any longer just about the West. As a proliferating thicket of acronyms suggests, new constellations of political power, economic growth, and demographic influence now cut across the old North-South divide: be it the BRICS (Brazil, Russia, India, China, and South Africa) or the PINCIs (Pakistan, India, Nigeria, China, and Indonesia). Between them the BRICS nations now account for 40% of the world's population and produce 25% of its output. Europe, by contrast, which has been accustomed to having a growing share of world population, now sees both its share of GDP and its population declining, and the influx of immigrants through which it seeks, against its own instincts, to address these demographic disadvantages creates only further tensions.

But for all that BRICS have shaken up the international donor landscape and the *economics* of global order, it is actually far from clear that they represent very much of a change from the Western approach to international *politics*. China has been even more active than Western agribusiness in land-grabbing across Africa, enclosing the African commons for its own extra-territorial gain just as surely as Britain enclosed its own commons in the past. China is also currently seeking to build another Panama Canal (this time across Nicaragua), and for largely the same reasons the United States took over the building of the first one. The BRICS more generally are increasingly prominent contributors to Western-style peacekeeping too, Brazil having led the UN Stabilisation Mission in Haiti. And, as Russia has made clear with respect to the crisis in Syria, it is clear that the BRICS also expect to exert an increasingly decisive political influence over countries in which they have strategic interests. Perhaps the greatest problem with today's preoccupation with "emerging nations", therefore, is that if we concern ourselves only with the relative motions of rise and fall, then we blind ourselves to the more fundamental continuities of a world within which the lives of rich and poor are constantly made in relation to one another.

What, then, in the context of all this, is global poverty today? A product of market failure? A global injustice? A curse of nature and environment?[38] And what, in turn, are any of these things to us? Of course, global poverty is in reality many things to many people, and if you listen to the moral philosophers, it is also nothing to far too many of us. But in truth there is no such thing as "global" poverty beyond what it is made to be in relation to the interests of the privileged and the powerful. There *is* great poverty in our world, but that poverty derives from the way that structural forms of injustice are made, over and again, through specifiable policies and traceable political decisions. Poverty takes its place as one of those critical things whose "right disposition" is insisted upon because doing so enables the whole modern complex of power that revolves around the interests of the rich to be overlooked.[39]

Power is everywhere, and we must be attentive to where and how it is used. But for all that many would have us believe otherwise, we live in a world of actually rather shambolic power and distinctly unruly leviathans. The "heath" has at best been only partially turned into a commons; worse, it has been pre-emptively and unevenly enclosed. We have a patchwork of international regulations and laws and authorities that abet power at least as often as they constrain it. We have a world of nation-states not so much jealously competing with one another as working out their position on the food chain: for while the power of some nations' authority is a matter of their own executive decision, other sovereigns cower deep within their war-torn shells, abdicating all reasonable public duties.

The solution to this is neither to embrace the wishful cosmopolitan hope of an ideal-type of "global democracy" nor to keep fretting about the state of isolated sovereign crises. The solution is to build up state capacity wherever global public challenges require this. The re-democratisation of the national state and the democratisation of the international must go hand in hand. We must recognise that the demons that confront us internationally (the mantra of security, for example) are not impediments to this task, nor reasons to run the other way, so much as they are the shadows of petty spectres we have ourselves created. We must look to the international domain not with fear or loathing but with a caring yet dispassionate eye, rejecting the salve of a philanthropy that seeks only to secure the conditions of its own comfortable existence, but in a disciplined manner, which is to say to the extent that we replace it with something that is better.

And the best way of doing this will be to devise new "strategies of equality", as Tawney once put it. These will not be strategies that seek the greatest quantitative equality in the shortest possible time; we cannot avoid the need to keep asking "equality of what?"[40] But we do need to develop a series of concrete and comprehensive solutions to the myriad political forms of inequality that confront us. And it is to a brief sketch of this task that we now turn.

5

A NEW EQUALITY

Among the laws that rule human societies, there is one which seems to be more precise and clear than all the others. If men are to remain civilised or to become so, the art of associating together must grow and improve in the same ratio as the equality of conditions is increased.
—Alexis de Tocqueville, *Democracy in America*

When forty-four nations, led by Britain and America, met at the New Hampshire retreat of Bretton Woods in 1944, there emerged from amongst the agendas, the intrigue, and the backdrop of war a commitment to creating a stable and orderly postwar international system. This was before the United Nations, before decolonisation in most cases, and before the Cold War came to dominate international politics. It was not especially representative of the world beyond the North Atlantic, and the vision of humanity it professed—a universal sanctity of human rights—was a vision frequently honoured in the breach. But trussed into place by wartime pragmatism and an appreciation of the value of mutual assistance, the underlying commitment to establish a binding form of international economic and political cooperation held up remarkably well.

A new international economic and political consensus is required again today. But how are we to fashion it? Comprehensive change tends to be made in the statutes of peace treaties alone. The Concert of Europe was born of the Napoleonic Wars, the League of Nations emerged from the mud of Flanders, and the United Nations and Bretton Woods were themselves products of the Second World War. But these were all diplomatic conversations between, for the most part, the very elites whose whims of national fancy had led their countries to war in the first place. The nearest thing we have to an international consensus today, the post–Bretton Woods doctrine of laissez-faire, market individualism has equally been an elite-driven project: a class war this time, waged on behalf of an emergent global 1%, at a time when the left was too hobbled, or too shamed, to resist it. We are familiar now with some of the results.

But perhaps a more democratic consensus does not require the folly of war to inspire it? A rolling, persistent wave of "defiant publics"—from the World Trade Organisation protests in Seattle in 1999 and Genoa in 2001 to food riots in Africa in 2007, and through to student strikes in Chile in 2011 and to Occupy and Indignado camps and sit-ins and lock-outs everywhere from downtown Seoul to Seville's Plaza Encarnación—suggests that we may indeed take hope. It certainly suggests that people the world over are beginning to find their voice. But finding a voice is only half of it, in short. Once found, you then need somebody to talk to. And while rich and poor had once to at least confront one another in time and space, today plutocrats and the precariat rarely do.

The situation today is different, then, from the way that it was diagnosed by Karl Polanyi, as he wrote about the rich countries the last time they were as unequal as they are today (during the interwar period). But his central observation, that it is wrong to believe that the demands of the market float above all other obligations—community, religion, politics—and dangerous to boot, holds still. The idea that market value can and should be the denominator of all else has been dominant again these past few decades. And just as Polanyi noted, it has led us to once again commodify life and thought, to overlook the use value of things in preference for their exchange value, and it has pushed societies close to the precipice once more.[1]

More hopefully, Polanyi also observed, there comes a point—the "double movement", he called it—at which society snaps back to its senses, and it may just be that this is where we find ourselves today. Of course, historically this return to common sense was the prerogative of the Western working class, mobilising to protest the fact that their own conditions of existence were being squeezed out from under them while others grew fat from the proceeds. But sooner or later, everyone finds they have some incentive for changing the status quo, and marching isn't always the way of it. When periodic downturns threaten the life of the economy, business groups will insist on changes to central banking to protect their market positions; when peasants can neither sell their own grain nor obtain sufficient other foodstuffs, they will insist on changes to national food-provisioning systems. When people refuse to accept that markets ultimately determine everything but that they do connect us all, then the extent of our own engagement with them is transformed into a constituency for change. When we cast our regard forwards and not backwards, our common fate, for richer and for poorer, is as one.

We are gradually waking up to this: not just the indignant and the poor in the rich world, who are tired of being lectured at and ignored by the establishment, but also the poor world's middle classes, who wish to ensure that their own recent exit from poverty is a lasting one, and the rich world's middle classes, who themselves now feel the pull into the drop that lies between the world's privileged and its poor. So too are a growing number of the very rich, who are tired of the felt need for security from the rest, tired of the "social cost" imposed by the misery of others, and who appreciate the fact that a world that excludes the majority of its citizens is a world diminished rather than preserved.[2] We all value and cherish creativity, fun, and freedom of choice. It is irksome for all of us to live within the confines of what "security reasons" prescribe. So perhaps we are in luck after all: it seems that we can proceed directly to the peace table.

A great many issues need discussing there. Above all, we need to speak more ambitiously than we like to any more about how we might actually address the underlying structural injustices we have allowed to lock into place. We must convert our era's obsession with international *economic development* into a parallel discussion about international *political development.* If the society we envisage for our children is to involve more than just a few good Samaritans plying their care in a charitable fugue, if it is to be driven by more than the elective two hoots of Internet slactivism, then it requires some form of engaged democratic and institutionalised politics—what Sheri Berman calls "the primacy of politics"*—refitted for global times. We must dare to imagine this, before forces beyond our control do it for us.

But democratic politics (as opposed to politicking) is precisely what has always been kept out of major international fora like the Millen-

*Sheri Berman, *The Primacy of Politics: Social Democracy and Making of Europe's Twentieth Century* (Cambridge: Cambridge University Press).

nium Development Goals, just as surely as it is kept out of the Security Council horse trades and the back-room deals underpinning the post–Washington Consensus. A global funding gap estimated at $180 billion every year says what we really thought about the MDGs; the funding gap for the Sustainable Development Goals, which from 2015 will replace them, is already estimated to be much greater.[3]

A lot of time, money, and public attention have been spent on mapping out this "new" SDG path to development through 2030. But the end result is little more than a shopping list of hoped-for ends (better health for so many billions, only so many deaths by malaria). The numbers are little more than best guesses projected into the future, and they distract from the fact that the underlying political dynamics remain the same. The goal of reducing tax evasion, for example, is slotted away under "strengthening domestic resource mobilisation", which is to say it is still thought of as the problem of Mobutu (and he died in 1997). Corporate tax evasion, rich countries' complicity in tax avoidance, and their arms sales to poor countries in danger of civil war all go unmentioned. In the absence of a wider political framework within which the underlying issues can be raised and tackled at the appropriate scale, powerful countries and communities will lobby and press their way towards a bewildering array of self-serving targets that will, almost by definition, fail to achieve their real goals. This is not the peace table that we need so much as a feast of the barons. We need something else entirely.

History offers us some guidance here—albeit not the tiresomely repeated history of how the West apparently managed to get things right, nor a history that privileges one scale of action (the global, the local, or the regional) against all the others. What is needed is a politics fashioned from the lessons of the past and directed at the challenges of today. It must instinctively be one of informed compromise, driven to implement profound changes that nonetheless stand a chance of becoming *real* change (what the clear-eyed economist Albert Hirschman once called "a bias for hope").[4] It must be a politics that prioritises (it helps to avoid such loaded terms as "delivers") social prosperity over economic growth. And it must be one that enables a deepening, not a hollowing out, of democracy. All of which means that there is, after all, somewhere we might begin to look for ideas. We shall make a leap to that (rather specific) place now, before working back towards what are our more immediately relevant concerns in the sections that follow.

"A BIAS FOR HOPE"

Social democracy is not a political vernacular that people tend to reach for when thinking of *international* politics. As a political project it is presumed to be confined to a few already-privileged and rather small nations. Its history is far more diverse than this, of course—and yet there are certain core features it is worth us bearing in mind. Above all, and in distinction to its two modern counterparts of state communism and liberal democratic capitalism, it neither was "built in one state" nor has been blindly and ceaselessly "exported abroad". During the Cold War years, communism and capitalism each sought to claim the world for themselves. Social democracy, by contrast, developed largely organically, in many cases in response to specific local concerns, and its strength has always come from not having a predetermined formula to address those concerns. Rather, it adheres to basic democratic principles but is willing to enact sufficiently universal solutions in their name.

Social democratic thinking has never impressed the right; it has never been the first to market—often, of course, for very good reason. It was traditionally viewed with suspicion by the left too, which often saw in it little more than just a long spoon for supping with the devil of moneyed interests. Today, as European nations see the remnants of their public sectors attacked in round after round of austerity belt-tightening, as inequality returns even to such deeply social democratic nations as Sweden, the values of the social democratic agenda are in danger of being forgotten altogether.

It is in precisely this context of crisis that the social democratic project needs inventing anew, at home *and* abroad. It is hard to see how else the world's deeply and differently vested interests will come together to act on the extent of inequality we see around us other than via a politics of compromise. Yet social democrats the world over will nonetheless need to think beyond the borders of the nation-state to rise to this challenge. This doesn't mean neglecting our own troubles as we counter those today; far from it. But if the social democratic gains of the twentieth century are to be in any way salvaged for the twenty-first, then they will require the support of a new wave of like-minded leaders and publics emanating from, and including, what are at present considered much poorer and politically less promising countries.

This requires a difficult admixture of humility and determination. Humility first of all: for if Western welfare capitalism has failed on its own terms, then it was never as good as we might like to recall. From this follows the need for wholesale reinvention. And while there is much to learn from the past here, including much that we have already forgotten, there is equally much to take on board from what is happening, and will be happening, elsewhere in the world. The challenge, as it confronts us today, is one of synthesis between these two.

But determination is equally necessary, since reinventing the social democratic project today, and the modern, international form of welfare capitalism that we need (and that can be realised only through it) will require this new generation of social democrats to first create and then control new international institutional levers—since that is where the drivers of inequality presently derive their freedom of movement. This may encompass everything from reform of the United Nations to experimentation with grassroots movements like Simpol, which channels widely held but hard-to-organise demands for more cooperative international policies onto the agenda of national politicians by means of coordinated individual voting (it is a form of democratic lobbying in which voter communitarianism replaces the power of money). It will involve issue-specific policy work-around, for example, progressive taxes, of the sort recently called for by Thomas Piketty. Together these social democrats must unhitch social democracy from its twentieth-century reliance upon distributing the fruits of high economic growth nationally and take the initiative of defining a twenty-first-century political project of lower, greener growth, of de-commodification, and of social emancipation.

Against such forward thinking, social democracy is usually put on the canvas today in a distinctly nostalgic impasto, a yearning for the values of the past clearly visible in the brushstrokes, most of them depicting mild and unassertive individuals in uniform tones. And yet social democracy has proved over time to be about the best means we have for addressing the crises that under-regulated capitalism repeatedly produces. Its credentials are well worth reappraising, therefore. It is true that social democracy shares with communism an inclusive and public-oriented vision of the future. But it listens to individuals, accords them their privacy, and works with modesty and reason in its answer to the question, what is to be done? At the same time, rather than bottle up old values and place them on a shelf, as conservatism does, it re-creates them in light of public dialogue. It is, for these reasons, a distinctly forward-looking politics tempered by the solid craft of pragmatism.

But what is it exactly that social democracy, or whatever such a project needs to be called today, has to teach us when it comes to the problems of *world* poverty, of uneven development, of global inequality? As the Norwegian academics Kristian Stokke and Olle Törnquist have pointed out, the answer is nothing much at all if we are interested only in the outcomes, that is, in what self-defining social democracies have become and whether we think it feasible that others might copy.* What matters, once we set aside the clichés of dun-coloured Volvos and long summers at the lakeside cabin, are the insights to be gleaned from the history of the *making* of the social democratic project, of the political dynamics involved, and of what these things can tell us about the conditions of existence for transformative democratic politics today.[5] And *that* history takes us back this time to before the Second World War (it takes us to other times and places too, but the following will at least serve as an example).

Scandinavia in the 1930s was, like most of Europe, plagued by crisis, industrial unrest, political radicalisation, and poverty—a situation familiar to many parts of the world today. Sweden was a "cauldron of conflict", in the words of one expert, and just thirty years earlier, in 1900, had been Europe's most unequal country. Governments came and went across Europe, in fact, as people lost faith in capitalism and then, increasingly, in the parliamentarians overseeing it. Norway and Sweden were no exception during this period: they both went through twelve prime ministers in fewer than fifteen years. What use was the vote when those you voted into power could do nothing to help you?[6]

*Kristian Stokke and Olle Törnquist, eds., *Democratization in the Global South: The Importance of Transformative Politics* (Basingstoke, UK: Palgrave Macmillan), 2013.

This was, of course, the beginning of the slide to fascism in many countries. And yet in Scandinavia, it was the beginnings of the social democratic project as a political force to be reckoned with. Unemployment was no less there, the agricultural crisis was just as severe, and racialised national ideologies—typified by the likes of Rudolf Kjéllen—were present. The difference was in the politics pursued by the parties themselves. Recognising that everyone was vulnerable to the economic crisis, the social democratic parties reached out beyond their core worker constituencies to the middle classes first, and then to the farmers. And they preached social unity as the means to holding a moderate political line in the face of external economic pressures.

This put the Scandinavian social democratic parties in a different position to others in Europe during the interwar period.[7] They offered sanctuary from the forces of economic and political destruction, and by defusing the tensions between those whose interests were variously (and quite differently) tied to capital, labour, or land, they substantiated that offer of sanctuary while also creating the political space necessary to come good on their own promises of modest improvements in material conditions. In the famous "cow trade" between workers and farmers in Sweden, for example, the Swedish Social Democratic Party (the Socialdemokratiske Arbetareparti, or SAP) accepted protectionist measures for some agricultural products in exchange for farmers accepting progressive labour policies. Compromises such as this helped underpin the social democratic era to come. They did so by giving people confidence in social democracy's public-spirited formula: by investing in collective needs first, citizens would reap a more reliable suite of personal benefits later.

The effect was not only social stability but economic prosperity too.[8] By increasing the competitiveness of the export industry, the social democrats created more jobs. By ensuring investment across sectors, they reduced the unevenness of the economy in general. It was good for workers (through pensions, social security, housing, education, and unemployment support) and good for employers (wage levels were based on what employers could pay given their international exposure, there was greater industrial peace, and the state took care of workers' social needs).[9] All of this created the conditions for more flexible and adaptable labour markets, including the take-up of hydropower—encapsulated by the vast Norsk Hydro station at Vemork, itself later immortalised in the 1965 film *The Heroes of Telemark*. And it gave a majority of people reason to support the democratic regulation of society regardless of who owned or controlled exactly what within it.

The dynamics were slightly different in each country. The social welfare reform that emerged out of the Kanslergade Agreement in Denmark in 1933 came at the height of the interwar economic crisis and was part of a wider compromise package to save the Danish economy. It was a coincidence, but a telling one, that on the same day that the agreement was ratified, Hitler was appointed chancellor of the Reich in Berlin, convinced that he had a better way of saving the economy.

In Sweden, the politics of grand compromise were pushed forward with the so-called Basic Agreement—essentially a commitment to the spirit of give-and-take in labour disputes—signed by the employer and labour union associations at the unprepossessing seaside retreat of Saltsjöbaden in 1938. "In fear of death one commits suicide," one disinclined wag at the Transport Workers Union is supposed to have said at the time. But in truth the Agreement was the foundation for the relative industrial harmony that would mark Scandinavian industrial relations throughout Europe's Golden Age.

In Norway, the empowerment of women as well as workers was key. Women workers marched in 1905 and achieved limited suffrage in 1907, with full suffrage coming in 1913, some five years before it came to Sweden. The need to build up an informed citizenry was recognised as equally important to women's suffrage and was met with the introduction of near-universal education policies at the end of the nineteenth century (very early in European terms).[10] That those policies were universal served another purpose, however: they enabled the government to appeal to those who lived in the far north of the country, who might otherwise be tempted to turn their allegiances away from Oslo, far to the south.

We are perhaps too cynical for this sort of grand policy-making today, and perhaps too wary of the paternalistic streak that can lie behind it. But health insurance (Denmark had 65% coverage by 1930), national pension plans (Sweden enacted the first in 1913), and unemployment insurance (Norway had 50% coverage by 1914) all helped bind those societies together. This insulated them from the divisive attractions of fascism, then quickening across the rest of Europe and provided the basis for economic prosperity down the line.[11] Supporting small-scale farmers when it was obvious that they were not the economic future also enabled more people to be ready for that future when it did arrive. Economically speaking, neither the laissez-faire policies of the nineteenth century nor the late twentieth-century workfarist reincarnation of the same have proved anything like as effective over the long term.

The Scandinavian countries thus each faced slightly different challenges and took different paths to resolving them. But in every case the achievement was two-fold. First, it enabled them to identify and develop a clear sense of national identity and collective responsibility, sufficient to override the economic, and soon political, crises of the era. This was achieved by expanding the range of politically enfranchised individuals and core interest groups, by affording greater legitimacy and capability to the state, and by offering everyone a reason to buy into that state in the first place. By using the state as a site of democratic contestation, rather than just a glorified butcher's hook on which to hang this or that predetermined policy, farmers were kept on the land and workers were kept at work—in contrast to elsewhere in Europe.

Second, these states made full use of what has since become a characteristic feature of Scandinavian social democracies: an emphasis on universal and preventive policies. The principal of universality distinguished these states from other approaches to welfare provision at the time, both in Europe (where the International Conference of National Unions of Mutual Benefit Societies and Sickness Insurance Funds was launched in Brussels in 1927) and America (which introduced its Social Security Act in 1935). Both of these were strongly means tested and targeted only certain groups of citizens (those of industrial cities in the case of the SSA).[12] But in Scandinavia, assistance was not about alms based on a calculation of diminished citizen status; assistance was a much more positive policy, intended "to have an integrative function, elevating the citizen to full citizenship".[13]

The commitment to more statutory forms of welfare provision was common to both sides of the political spectrum in Scandinavia. Sweden's national pension plan was put forward by the liberal-conservative government of Arvid Lindman as a ploy to retain domestic workers who were emigrating in ever-greater numbers to the United States. The political case for these policies, in short, was made up as these societies went along. Indeed, had the Scandinavian countries mapped out pre-emptively, in a series of five-year plans, what they would go on to achieve, the entire social democratic project would have been greeted with hoots of derision on both sides, because none of these countries was starting from a position of economic or social strength at all. Probably, it never would have happened. In 1900 half of the Norwegian town of Stavanger were still canning fish, and most of the Scandinavian countries remained largely agricultural into the 1930s. Up until 1910, Sweden was possibly the most indebted country in the world.[14]

And yet it soon became clear that a politics based upon the universal provision of social protections, the enfranchisement and incorporation into the political processes of previously under-represented groups, and the primacy of a politics of grand bargaining would not so much hinder prosperity and growth as much as be an eminently acceptable means of securing those.[15] A sense of quite what was achieved comes through in a diary entry by the Swedish prime minister Per Albin Hansson, written at the end of Sweden's "harvest season", a period from 1937 to 1939. It was 1937, Hansson began, that really "loosened things up":

> That is when a pension amendment indexed to the cost of living was enacted, child support, mothers' assistance, maternity assistance, far reaching improvements in preventive mother and child care, the housing loan fund. The regulation of farm labour was improved. 1938 gave us compulsory holidays, the national dental plan, and the Institute for Health Insurance. 1939 saw the regulation of working hours . . . [and] housing for pensioners was created for the aged.[16]

As he perhaps did not need to conclude: "In this period Sweden was ahead of Norway, and, indeed, most European countries."

But the other Nordic countries were not far off. Finland, which for some time had lagged behind Sweden, Denmark, and Norway, and which as a result had found it hard to mobilise capital for development before the war, released money from its tax-funded national pension scheme in the 1950s to build power stations and a national economic grid. Similarly, in Sweden, the proceeds from the world's first public pension scheme, which had been introduced in 1913, later provided housing for the new urban population that economic development (and agricultural decline) was drawing into the cities.

Scandinavian countries enjoy the benefits of all this today still, for all that their current crop of leaders is more concerned to coast on the earlier gains than to expand on them. For one thing, these countries continue to privilege a model of dual-income households, as compared to European and American welfare systems, which tend to assume a male breadwinner and a female homemaker, and there are economic and productivity gains to be had here. It has recently been estimated, for example, that if full-time female participation in the Norwegian labour force were reduced even just to that of the OECD average, the country's net national wealth would fall by a value equivalent to Norway's total petroleum wealth.[17]

For another, while mature social democracy in Scandinavia developed, historically speaking, into a political model that was largely about the management of industrial relations, the *political origins* of the wider social democratic project—and above all the constellations of social movements, political compromise, and social policy formation that it entailed—were formed by a broad array of interests brought into a democratising political order (not necessarily a fully democratic order).[18] In its broadest sense, then, social democracy was really about ensuring that social inclusion was always the tool with which to cut the cake of economic growth—or to share out the costs when the cake gets smaller.

From this flow several important observations. First, it is possible both to "democratize before democracy" and to provide "social assistance [that is, social policy] before the advent of the welfare state"—and perhaps even before a widespread commitment to it.[19] But both of these things rely upon affording people representation and a real voice in policy decisions, not the "empowerment" of so many failed policies in poorer countries and regions, and not the "trickle down" of so many failed policies in wealthier parts of the world: a trickle-down which, as Ann Pettifor points out, is invariably dwarfed by income from rents heading the other way.[20] Both also supply the conditions for economic prosperity in good times as well as bad. It is not by chance that social democratic countries today score highest on both quality of life and competitiveness rankings.

Second, it is apparent that this scale and type of political change is cumulative, not revolutionary. One cannot put all these pieces together in one stroke. Accordingly, there was not a single moment of compromise in Scandinavia's history but rather an emerging history of compromise. The September Compromise of 1899 in Denmark (between employers and employees), the December Compromise of 1906 in Sweden (between the Employers Association and the Confederation of Trade Unions, known as Landsorganisationen i Sverige, or LO), the Iron Agreement of 1907 in Norway (securing a minimum-wage principle in the metals industry): these came first. They were followed by a series of legal contributions, such as the Industrial Disputes Act of Norway in 1915, with its principle of compulsory mediation. Only then could the Main Agreements (in Norway in 1935 and at Saltsjöbaden, Sweden, in 1938) realistically take place.[21] In short, if inequality is made, not found, in the world, so too are the norms, values, and institutions with which it may be reduced.[22] But all this needs a longer-term political perspective to take effect.

Third, and related, is the importance of universalism. Universalism is one of the most effective ways of breaking the monopoly that dominant groups and classes hold over the policy-making process, at least if we wish to avoid subjecting them to Robespierre's form of political morality. Universalism underpins what Hannah Arendt usefully termed our "right to have rights".[23] If we want people to treat one another equally, then all must have equal opportunity to determine the rules by which they shall be governed. As the Danish social scientist Gøsta Esping-Andersen puts it, "The beauty of the social democratic strategy was that social policy would *also* result in power mobilization."[24] In short, cooperation became so important in Scandinavia not because conditions there were so good, but precisely because they were bad, because there was economic crisis, and because the answers to these problems were not at all obvious to any one person at all.

FROM ECONOMIC GROWTH TO SOCIAL POLICY

Social democracy is on one level, then, a sort of institutional sticking plaster: a contrivance beneath which the tissue of human solidarity can form, prior to its exposure to the market. After all, even Crusoe took with him to his island a store of accumulated nurture. It has become fashionable today to decry the social democratic "nanny state". But social democracy as a political project says nothing itself about the sorts of things people should do with their freedom. It merely wants to ensure that as many people as possible get to choose what to do with it. So the question is, why do we not see more of this around the world today?

The most common response to this question is to generalise on the basis of a geographical conceit. This is the view that social democracy is the product of small, culturally homogeneous northern European states and has little to do with the realities, above all, of poor nations. Yet the average population size of Latin America's nations (including the French overseas departments) is only around 23 million, whereas for Africa's fifty-six self-declared nations it is slightly less, at around 17 million. The population of Nigeria—the most populous African nation—is still less than half that of the United States. The population of the Democratic Republic of Congo is only a few million more than that of France; Sierra Leone's is around that of Norway's, South Sudan's that of Sweden, and Mauritania and Namibia's together about that of Denmark. The one truly vast capitalist country, India, is a federated polity, whose vast population of 1 billion has long been addressed primarily at the level of states (where the figures again resemble those of the European average).

A second common response is to claim that poorer countries simply can't afford the luxury of doing what wealthier nations do; therefore, they need to focus on being more competitive if they want to survive in today's global economy. But as we have seen, the Scandinavian countries were themselves riotously under-developed when social policies were first introduced there. Moreover, many countries have in fact tried some version or other of what the UN Research Institute for Social Development calls "welfare developmentalism", and they might have done more of it had they been allowed to. In Mauritius, for example, social spending is 40% of public expenditure, which is better than in many wealthier countries.[25] Meanwhile, over the past twenty-five years, Brazil has seen one professed social democrat (Fernando Cardoso) hold two terms before being replaced by a more radical version of the same ("Lula" da Silva) who was himself duly replaced, after two democratic terms, by a yet more radical version of the same (Dilma Rousseff). In short, it seems reasonably clear that large numbers of people in lower- and middle-income countries think that social democratic policies are precisely the sorts of policies they need. As the years of neoliberal ascendancy finally fade to black, their voices and experiences are again beginning to be heard.*

Even a no-nonsense liberal like John Podesta, Bill Clinton's onetime chief of staff and more recently counsellor to Barack Obama, has finally come around to this view. Social policy is essential, Podesta wrote in the *Guardian* in May 2013, not only for providing a "floor of protection" for the poor but also as a form of "cost-effective insurance" against the risks that confront us *all* in today's world.[26] In this article, he was speaking in his latest incarnation as the US representative to the High-Level Panel of Eminent Persons on the Post-Millennium agenda. To wit, he concluded: "We can end global poverty, but the methods might surprise you." They should not surprise us, however, since there have been times and places when those methods were taken for granted. Indeed, since the financial crisis, more and more poor countries have been increasing their public spending allocations (on things like health, housing, and education).[27] But if social policy is indeed the core of what we need to reduce inequality, then it is worth being clear on what exactly that might be taken to mean today.

"Social policy" refers to any number of systematic public interventions carried out in the name of protecting citizens from risks and vulnerabilities so that they are best enabled to realise their own particular ambitions.[28] It is a means of ensuring justice between people by promoting fairness in the distribution of resources (such as income) and

*Richard Sandbrook, Marc Edelman, Patrick Heller, and Judith Teichmark, *Social Democracy in the Global Periphery: Origins, Challenges, Prospects* (Cambridge: Cambridge University Press, 2007).

services (such as health and education). Social policy ought therefore to hold some considerable potential for rich nations and poor: securing a more sustainable form of economic growth by enhancing processes of democratic inclusion. As the former director of UNRISD Thandika Mkandawire points out, social policy can even be put to the task of nation-building, since it helps foster a sense of collective identity without the need to go to war. It demands of citizens that they become more peacefully versed in the habit of reciprocation.[29]

Social policy is both a useful and a progressive tool, then. Yet it needn't be something the left alone plants a flag upon, since it has frequently been used as a means of stabilising societies at times when the molten mass was seen to be getting just a little too liquid. It was Otto von Bismarck who introduced social policies to boost industrialisation and undermine the appeal of socialism in Prussia in the mid-nineteenth century. And that other conservative reformer, Eduard Taaffe, did the same in Austria to stave off what he saw as precipitate demands for democracy there.[30] These are not experiences to replicate, perhaps, but they are at least illuminating.

We stand to learn more, however, from the emergence of social protectionism elsewhere in the world. A long way from imperial Austria, social policy was an important part of South Korea's rapid national development in the post–Second World War era, in which experience private actors played an important role in the provision of social protections. Indeed, companies in East Asia more generally enabled the adoption of accident insurance protection for workers there, long before those countries had reached a level of development comparable to that of Europe.

South Korea also made clear the importance of land reform and the restructuring of society that it enabled, since it was the US-overseen land reform—modelled closely on earlier policies in postwar Japan—that broke down the structural basis of the old social order and fostered improved social mobility. As a result, poor peasant families were able to send their children to school to benefit from educational reforms, which itself played an important role in keeping income inequality down during the country's rapid economic growth in the 1980s.[31] This, then, apropos our earlier specification of the problem, was democracy before democracy.

In Taiwan, too, a limited form of welfare was provided by the major domestic corporations, by whom the majority of higher-skilled workers were employed, while the state operated as a sort of watchdog to the process.[32] It was an arrangement more than a system, and it left many workers uncovered, but it at least had the effect of embedding a culture of social protection into the heart of national economic policy-making. Here, then, was welfare before welfare.

Even in Africa social spending as part of a series of "national questions" was pursued with some vigour in the 1960s. This was a decade during which the level of economic growth across much of the continent permitted a certain reinvestment of proceeds. It was also a decade inaugurated by the murder of Patrice Lumumba in 1961 and so a decade in which social welfarism in Africa was not fated to last. During his short time in power, Lumumba had led his newly independent nation of the Congo with a strong national commitment to his people as a whole, and not to one ethnic group in particular. Following Lumumba's murder, however, and the bloody international squabble over the mineral rich Congolese region of Katanga that ensued, the country was walked a step in precisely the wrong direction. So too were would-be democrats elsewhere in the continent given an object lesson in the fact that democracy, for them, was not going to pay.[33]

Given the intense international interest over the Congo and its resources, its subsequent descent into a pro-Western puppet state under Joseph Mobutu was always something of a special case, perhaps. But the Congo proves a more general rule too: that the moment social spending—one of the few tools with which disparate ethnic and cultural groups could be bound to a national rather than ethnic community—was replaced by spending on armed force and security, the Congo's people found themselves one step further removed from any potential returns on their resources.[34] This is partly why, even today, for every percentage point increase in gross domestic product, less riven nations like Malaysia and Vietnam have reduced poverty ten times faster than African countries like Tanzania.[35]

It was in Latin America, though, from as early as the 1930s and at a time of recession not all that removed from the Great Depression, that something more clearly approaching a "developmental welfare state" most clearly came into being.[36] Mexico began introducing social provisions such as public education in the 1930s. Pension programmes for urban workers had been introduced in the nineteenth century in Argentina and Uruguay. And in Costa Rica, which is today one of the highest-ranking Latin American countries across a range of indicators, social policy was a central part of the country's development strategy from the 1950s onwards.

To be sure, Latin American nations were at an advantage here: not least, their borders had been less artificially and less recently imposed upon them than was the case in Africa or East Asia. In Latin America state boundaries were more clearly "national" in the Western sense of the term and allegiances to said state were accordingly more "natural". This made the task of national social spending easier. As did the existence of the regionally focused Economic Commission for Latin America and Caribbean, which always refused to believe in the idea of standardised "stages of development" and whose research and policy advice were informed by an underlying Keynesianism. But each of these cases still serves to highlight a basic point: the political process of social policy provision as a central part of more democratically minded economic development strategies is every bit as fundamental to creating the conditions for social prosperity as are the property-endowing institutions beloved of Daron Acemoglu and James Robinson, of the aid beloved by Jeffrey Sachs, and of the market freedoms championed by William Easterly.

Nowhere has this been clearer than with Brazil's attempt to scale up to the national level the kind of democratic decision-making over public policy that is usually seen at the local or municipal level. Brazil provides us with an important series of political lessons in what can happen when basic, universal forms of social policy are given their due. Simply by ensuring an effective universal pension scheme, for example, Brazil moved 18 million people out of poverty in 2001.[37] It has since reduced the percentage of the population living in extreme poverty from 14% to 4%, in just ten years.[38]

The Brazilian experience has limits, of course. Its much-hyped cash-transfer programme—the Bolsa Família—is a targeted rather than universal programme, and while the government has given people a greater say in matters of social policy, it has not always given them the money needed to realise those ambitions. But what we are interested in here, as with the Scandinavian experience of the 1930s, is less the substance and achievement of any one particular initiative than the political process through which they came about. Long before Lula, before Brazil was an emerging power even, the now-governing Partido dos Trabalhadores began its own rise to power as a force in *regional* government. The party's success lay with its ability and willingness to incorporate social movements and actors from outside the party. This became known as *o modo petista de governar* (the PT way of governing)—an approach to government that unconsciously echoed the policies of the Swedish SAP or the Norwegian AP in the 1920s and 1930s.[39]

The PT's efforts to reach out to the people were complemented by a rise in the popularity of social movements coming the other way: most famously, from 1989 onwards, with the participatory budgeting forums for municipal planning established in Porto Alegre (in which citizens met in various year-round assemblies to identify priorities as they saw them and to allocate a certain portion of municipal funds accordingly). This same ethos of participative policy-making came to be increasingly applied at the national level as well, leading to the creation of Brazil's Unified Health System in 1990. Many of Brazil's later social and economic successes—despite the country's ongoing problems—have similarly been based upon the gradual building up of this political platform and the growing trust (in a post-authoritarian period, lest we forget) it has enabled between citizens and the states.[40]

Costa Rica is another case in point. One of the poorest countries in Central America prior to the Second World War, today the country is sometimes described as the Switzerland of Central America. The turning point was set in train in the 1950s, when the National Production Council and National Wage Council were established to ensure that grains needed for a basic diet were affordable for citizens and that minimum-wage floors were set in place in different industries. The National Housing and Planning Institute built dwellings and provided subsidised mortgages for people to purchase them. Meanwhile, a road-building programme doubled the length of the nation's highways.[41]

The result was a mini–golden age, with Costa Rica posting some of the highest growth rates in the continent along with an "exceptional" record on social progress. Poverty and inequality were reduced. And this was despite oil shocks, despite the country being temporarily cut off from some of its major trading partners by a war between its neighbours (replicating without succumbing to the presumed major problem of landlocked nation status) and despite external aid in the form of US Agency for International Development payments and "advice" from international financial institutions undermining the model somewhat in the 1980s and 1990s.[42]

The Indian state of Kerala offers yet another example. Kerala is not rich. It struggles with a per capita income of $500 per year. And yet it has achieved a 93% literacy rate and a life expectancy of seventy-two years. It is also the Indian state that has seen the greatest poverty reduction in recent decades. Why? The answer is not a sudden upswing in economic growth or investment but has a lot to do, rather, with the extension of social welfare to all, including the many informal workers, regardless of ethnicity and caste. The result is a state where violence is substantially lower than elsewhere in India, where electoral participation is between 15% and 20% higher than in the rest of India, where what was once India's most rigidly enforced caste system has been shelved (officially at least), and where Muslims and Christians take part in elections together.[43]

If this is promising, it is also no more than what Gunnar Myrdal foresaw in his 1974 Nobel Prize acceptance lecture, "The Equality Issue in World Development". He followed that up in 1978 with a lecture in Saltsjöbaden, appropriately enough, on "the need for [social] reforms in developing countries".[44] Myrdal saw that reform of both the state and the market in poor countries was necessary in addition to aid. That perhaps hardly distinguishes him from the Washington Consensus train of thought. But having himself been one of the key thinkers behind Sweden's postwar economy, he recognised that politics too must be reformed in such a way as to control the market: more equality would result in greater growth, and political interventions like land reform would be required to get to greater equality first.

Societies need to find ways to protect themselves from themselves, in short. We are none of us perfect decision-makers, and we should not be expected to be sensible every day of our life: for all that the champions of "poor economics" celebrate our innate wisdom and market savvy, sometimes it's nice to have something to fall back upon. Myrdal would likely have looked with favour upon recent developments in the Philippines, for example, where public financial management reforms have given a once subjugated civil society oversight of the budget cycle (driven this time largely by non-governmental organisations).[45]

What is necessary, then, is not that we tell poor countries how to develop, as the World Bank and the International Monetary Fund and the Millennium Development Goals have all developed a comparative advantage in doing, but that we encourage their efforts towards reform in light of their own particular needs and historical experiences. That should involve progressive tax policies (taxing wealth and incomes, not goods, and ensuring that the proceeds are spent to mobilise the capabilities of the poor), government intervention (with active states promoting employment, regulating markets, and opening themselves up to civil society as they arbitrate between classes over conflicting interests); it should involve redistribution (with welfare states working on the principal of universality, or as near as possible to it, so as to protect citizens against loss of income, ill health, and loss of land); and it should involve active social policies (countries need not just schools but also school *systems*, and a commitment to training and paying the teachers to work in those systems).

None of these policies derives from development. They are how countries get *to* development. They are also how rich countries remain in development. But this requires remembering some of the more successful experiences of post–Second World War national development, before they were shaken by the political earthquakes of the 1970s and turned over into ruins by the neoliberal consensus we have toiled under ever since.

Such policies are more nearly within our reach than we might think. But to speak of such things is not, to avoid a likely misunderstanding, to raise the spectre of "global social democracy". The examples here are of primarily national and subnational achievements. They offer no generalizable rules, and it is far from clear that we should try to emulate them at the global scale, even if they did (this for the same reason we do not need to be trying to get back to some presumed pristine version of the Western welfare state: there is a reason it is in crisis). They do, however, offer a wealth of insights: above all regarding the nature of the political innovations required to underpin social policy approaches to development.

It *does* seem reasonable, therefore, and as former Brazilian president Fernando Henrique Cardoso once put it, to speak of "global social democracies", that is, states with social democratic systems that are in tune with global realities and that are not just protectionist enclaves, states that are, in that sense, *socially* globalised. Such global social democracies are a chimera in today's world, of course. But they give us a shape and form to work towards: not anti-globalisation but not uncritically "for" globalisation, either. They recognise that the outcomes of their hard-won social policies offer certain advantages in a highly competitive world: skilled and healthy labour forces, quality infrastructure, orderly industrial relations.[46] They recognise that the question is really, which globalisation do you want? And they know, from experience, that the answer is "Something you can give yourself a say in". Individuals cannot change globalisation directly. But by changing the nature of the policy regimes that manage it, we can at least keep it from becoming those things we don't want it to be.

There are a good many obstacles confronting poorer countries in this task, many of them historically grounded, as we have seen. Often a country has had social policy, but of such a scant or crooked variety that the idea itself is tainted: colonial-era welfare, say, that either benefited the governing classes alone or was used to curry favour domestically, if not to entrench existing fault lines between different races and ethnic groups. In some cases, as in India, there are simply not enough funds available for nationwide social spending at present, and this is in no small measure because not enough citizens pay their taxes yet. In Liberia the money for social spending *is* collected, but it goes towards paying off the debts previously incurred by local elites. For many countries, further questions arise from the fact that certain structural conditions for social policy delivery (and social democratic politics as a means of achieving that delivery) either do not exist or exist in highly modified form. To take India again, for example, there is not so much a unified working class as a vast and largely unregulated "informal" labour force.

There are a great many obstacles confronting us in rich countries too. But the answer is not to give up hope in light of these challenges, or to believe that we are best off fixing them in isolation, but to take note of three core lessons from the social democratic project of the twentieth century—compromise, distributive fairness, and democratic deepening—and to find ways of reinventing them for the global age. This is no place for details, but we could do far worse than to begin trying to imagine how each of these might look, beginning with the political art of grand compromise.

A JUST MEASURE OF MORALITY

A much greater gift than sympathy that today's rich nations could give the poor world would be to use the leverage we have over our own states and corporations, and the greater say that we have in the international institutions which govern the relations between them, so as to even up the global playing field. Tax loopholes are an obvious place to start. But closing them is not quite as simple as boarding up the thousands of tax havens around the world as if they were moonshine joints. Many tax havens are sovereign states themselves, like the Cayman Islands, whose only comparative advantage is, in effect, its competitively low taxation. Closing these loopholes requires a globally consistent way of regulating the corporations who place their profits there.

But this in turn requires that our own governments be convinced that we, as rich-country citizens, actually care about any of this—that we think it is of some importance whether they are to act or not. "You can't just point at things and tax them", said the pop star Myleene Klass, in a famous television exchange with the leader of the British Labour Party. And she is quite right, of course. But some things are so obvious they need no pointing at. They just need taxing. To wit, the current failure to regulate global tax loopholes is undermining the Exchequer of rich countries as well, to the tune of around $13 trillion a year.[47] This makes it all the more inexplicable that in 2012 the British Treasury introduced yet another tax loophole—a relaxation of "controlled foreign company" rules—that makes it easier for British multinational companies to avoid paying taxes both in the poor countries where they earn their profits and at home.[48] So we are naïve to rely on our governments alone to sort this out. International rules are needed too. But this can't be done in one step. The Organisation for Economic Co-operation and Development has launched an action plan on "base erosion and profit shifting". The African Union needs one too.[49] We must build from there.

But as often as not, what is needed are not new rules so much as a proper application of existing ones—to keep pace with the widening scope and scale of financial activity, for example. Whenever interest rates slacken off in rich countries, a vast amount of capital takes shelter in emerging economies, usually for just a short while. This flood of digital money crashes domestic exchange rates and wreaks a distinctly more tangible havoc: it makes local industry uncompetitive, it leads to hikes in food prices, and it undermines (ironically, given the source of these problems) investor confidence in the country.

Poorer countries, then, should be afforded greater power to control such volatile capital flows across their territory. Brazil, Argentina, and Costa Rica have all experimented with capital account regulations since 2008, the better to enable them to react to price volatility and to give their governments the room for manoeuvre they need to make such monetary policy decisions as their national economy requires.[50] But they require international support for more fully fledged regulation of the global monetary system to succeed. As it stands, financial interests and richer nations reject these policies as "protectionist".[51] When Dilma Rousseff asked David Cameron and Barack Obama to limit capital flows coming from their countries into Brazil (a product of low interest rates and slow growth in the North), because they were making it harder for Brazilian firms to export and jobs were being lost, both refused, leaving the Brazilian finance minister to scramble to put together an overnight response (the best he could do was impose a modest 2% tax on foreign purchases of stocks and bonds).[52] The world is indeed not flat, for all that some would make it so.*

*Thomas L. Friedman, *The World Is Flat: A Brief History of the Twenty-First Century* (New York: Farrar, Straus & Giroux, 2005).

The regulation of capital account flows is central to securing a more stable and just international order: currency wars cannot be the way of it. Yet we are presently moving in the opposite direction. Rich countries today pre-emptively seek to resist *all* forms of regulation on new transactional forms, just as they have continually sought to limit the development of capital exchange controls since the 1970s. The IMF's grudging acceptance of "last resort" measures, which are supposed to address this, is scarcely worth the paper it is written on. This is short-sighted as much as anything else: the reason for low interest rates in the rich countries in the first place was to try to encourage growth and productivity there, not speculation abroad.[53]

Again and again we are told that this is just how the global economy is. Yet when we look at things historically, we know this is not true. We know that money's freedom to move across borders is historically very recent (just as present-day tax rates are historically extremely low).[54] We are told that such regulatory measures that, even the IMF acknowledges, could improve global welfare would raise "global costs". But that is only partially true as well: such measures would raise costs only for certain financial interests: almost all the rest of us would benefit. Such claims of incapacity by Western state and market powers are nothing if not disingenuous.

Disingenuousness is a habit of ours that we would do well to take note of and address. As we have seen, for all that Alan Greenspan—along with Bill Clinton and Tony Blair—painted the 1997–1998 financial crisis as being Asia's "fault" (and successfully, it might be added: we remember it even today as the "Asian crisis"), it was in fact largely a Western private-sector debt crisis that exploded into the public domain in Asia. Yet if it wasn't Asian in its origins it was nonetheless Asian in its effects. In Thailand, for example, it resulted in an economic contraction of 11% in 1998 and a 40% increase in the poverty rate.[55] But Western financial interests—and the governments of the United Kingdom and the United States in particular—that were then, as now, best positioned to do something about this instead chose to do nothing at all. And so it spread to other parts of the world as well. There simply isn't anywhere else to turn here. As we might say, modifying just a little that apocryphal quote attributed to many an American president, "The modern finance industry may be a son of a bitch, but it is *our* son of a bitch."

There are certain areas, then, in which rich nations have not just a special responsibility to act but a far greater capacity to act too. And for all the talk of empowering the citizens of the poor world, it would be helpful also if the citizens of rich countries had a more direct say in such international rules and regulations as affect us all. It turns out, for example, that most citizens of rich nations are more in favour of debt write-offs for unjust debt burdens in poor nations than their leaders are, and with good reason too: in those countries that have qualified for debt relief, primary school enrolment has increased from 63% to 83%; in countries that have not, like Sri Lanka and El Salvador, more than 25% of the national budget still goes towards debt payments.[56]

Getting our own house in order is a first step, then. But we need to go further. We need to address, in all seriousness, the welfarist challenge as it was first raised most directly by the Cambridge economist A. C. Pigou just over a century ago: that of increasing the share of income going to the poor (the question of distributional fairness) and of reducing the variability of income itself (the question of macroeconomic stability), a challenge it now behoves us to solve internationally.* Economic welfare is but a part of welfare more broadly, as Pigou himself recognised, and as we are by now familiar with in rather more concrete terms. Inequality too is about more than income inequality. On both counts the need for a form of "global public economics", as Tony Atkinson calls it, has never been greater: and so it would seem a good time to speculate a little as to just what, exactly, that might involve.[57]

GLOBAL PUBLIC INVESTMENT: A CASE STUDY AND A STEP IN THE RIGHT DIRECTION

What if we were to take the present impetus and rationale of the aid industry and transform it into a system of global public investment: a system with the potential to benefit us all rather than just a select group of means-tested recipient countries?[58] In some senses this is not as

*A. C. Pigou, *Wealth and Welfare* (London: Macmillan, 1912).

novel an idea as it may sound. The Marshall Plan was an act of international redistribution writ large from America to Cold War Europe, and one from which most Europeans today are just the latest generation of inheriting beneficiaries. As it carries that baton forwards, the manner in which the European Union today distributes resources amongst its quite variously developed regions, reaching individuals and communities in parallel to the policies of national states, is even closer to a working example of how public spending across national boundaries might work.

While falling some way short of the sort of "global welfare" that the right will cry "Foul!" and "Impossible!" about at one and the same time, global public investment may be best thought of as a form of distributional spending that actually seeks to avoid the need for constant aid handouts, which are resented not only by the right but also by many of those receiving the aid. More specifically, global public investment would collect together a range of funds derived from various wealth-creating activities and places, and then channel them into longer-term, managed investments in basic public goods and services, *wherever those goods and services are needed*. It would be primarily an intergovernmental initiative, with the overall level of finance to be raised set independently (though nation-states themselves would be able to determine the actual tax base), and it could take various forms: a tax levied on entities, groups, and institutions, for example—though individual tax receipts could be factored in as well.

Within nations, public spending of this sort is entirely uncontroversial. It is widely accepted that governments will redistribute resources so as to channel resources to regions and individuals in greater need. For example, Brits don't ask (too many) questions when the government invests in public infrastructure in "the regions", instead of, say, the South-East. Norwegians don't give much thought to the fact that their government sends trainee doctors to the round-the-clock winter darkness of Alta and Kirkenes; in a "free market", the Finnmark region would otherwise have very few doctors indeed. We accept that much of what we pay in taxes will go to people and places we will never see, and that, in fact, is one of the basic objectives of taxation: to sustain a collective greater good beyond the scale of family, kinship, or community. It also serves to impose discipline and to foster a greater procedural transparency. Modern nation-states simply could not exist without it. As is becoming increasingly clear, nor can the world either.

We pay taxes in any case not just to help others but also to help ourselves: by making a moderate contribution to a public body we create the means for those authorities (usually states) to provide us with such things as we ourselves have neither means nor desire to provide—be it transport infrastructure, new technology, or public health and education for our children. The success of national taxation schemes rests upon a judicious balance of public and private interest. The same is true internationally, which is why a system of global public investment would not be about some countries merely subsidising others, and for two important reasons.

First, countries that would be primary recipients of global public investment would *themselves* contribute something to the pot. This is on the dual grounds that with such a system the mix of givers and takers will be far more fluid than it is today while, in the meantime, the act of contributing, and the institutional discipline that requires will be useful in building up infrastructure and trust: two things sorely lacking at present internationally. Second, it is wrong to conceive of such resource transfers as only about respective national governments disbursing money between themselves at the national to national level: global public spending could work across regions and between different groups globally as well. It has the potential to be as flexible as institutional capacity allows.

The question then arises of just exactly how to fund such a scheme. To begin with, it could be as an additional component of existing taxation schemes, for example. Even just ensuring that all governments simply meet the long-agreed-upon 0.7% target for international aid contributions, for instance, would open up considerable funds, beyond those we presently have to work with. Funding would also of course be something met by *all* countries, substantially raising the tax base as it does away with a binary world of donors and recipients altogether. The global public investment model recognises, then, that in some senses “we are all developing" still, as Jonathan Glennie—one of those to have most convincingly outlined the case for such a form of spending—argues.[59]

This faculty is precisely what would make GPI more effective than aid: for it would provide a means to strategise, to think forwards and not just backwards in a fog of reparation and patchwork relief. It would resurrect the best of the five-year plan and combine that with the hard-won lessons of half a century of aid. Perhaps above all, such an ability to use a large pot of money to address the underlying problems that poorer parts of the world confront is precisely what makes GPI amenable to factoring social policy goals into questions of global development.

Social policy in the GPI model could potentially result in what some refer to as a "double dividend". Just as we levy taxes on products that are harmful to our personal health, so could global public investment be levied at one end, at a fractional rate, upon economic activity that directly jeopardises obviously public goods such as health.[60] At the other end, its disbursements could be prioritised for investments in things such as building infrastructure, where needed, or emergency food provision. In short, GPI has the scope to be infinitely more flexible than aid, which is always beholden to conditionality and the whims of literally thousands of donor bodies, all working in competition with one another, for all that they share the same broad objectives.

There are already a growing number of practical suggestions for some form of global public investment, including suggestions to tax a range of financial transactions. This includes the Tobin tax, which would take a sliver from financial transactions in foreign exchange markets, which currently turn over $4 trillion each day but are not subject to any meaningful taxation. At a proposed rate of one-two-hundredth of a percent, a tax on these financial transactions could still bring in more than $30 billion a year.[61] Levies of this sort could raise considerable capital resource, which would partly feed back into regulating the industries whence the money comes and partly feed forwards into a larger social pot.

Banks and corporations will protest when the rules of the game are changed, of course, but they must be reminded that they have already enjoyed their own private "golden age" of under-regulation and under-taxation around the world, price fixing within their own corporate structures and playing the global economic checkerboard to full advantage. Meanwhile the old kings of state are stuck fast, sovereigns of nothing more than their own territorial squares. It is time for that relationship to change.

Some, perhaps even most, governments will oppose such rule changes too, as Britain's present chancellor of the Exchequer George Osborne did when he sought to challenge a European financial transactions tax because it would harm the interests of the city of London. But here too there is a case to be made. For as was pointed out to him at the time, more than half the current £3 billion in revenue that comes from Britain's existing tax of this sort, a stamp duty on stock sales, is revenue coming into Britain from *foreign* entities.[62] There are indeed benefits to be had by all.

This brings us finally to the question of management: How would the money be collected? Who would decide where it would go? How would it actually get there? A global registry of wealth would be needed to start with.[63] With global public investment, transparency would be even more important than it is for aid at present. But these are not novel or insuperable problems: our own national taxation systems find ways of addressing them to an acceptable degree. The collection, management, and disbursement of such large sums of money internationally would present institutional challenges beyond the national scale too, of course; although as Tony Atkinson, suggests, the hurdle to actually starting something like this up could be overcome by means of a form of "flexible geometry", starting with the European Union, for example.[64] So these ought not to be insuperable problems: indeed, they must be overcome. The current tax regime is one built upon the principles of a territorially enclosed era, and yet we live in a globalised era: nations compete more forcefully through tax policies today than they do via tariffs and trade policy.[65] But that still leaves the question of accountability.

The United Nations has for decades cultivated a public trustworthiness and, for the most part, impartiality as an international body. Only the United Nations could provide a sufficiently accepted and accountable umbrella under which a UN Public Investment Authority, say, could be established: its headquarters in a large international city outside of Europe or America, somewhere like Tokyo, perhaps, but with regional public investment branches of the organisation too. These regional branches would ensure more local accountability, confidence, transparency, and competence in disbursement. There would be costs involved in setting up anything like this of course, but the costs are marginal compared to the benefits of what such an organisation might bring.

But what, then, would GPI actually do? Why go to all this trouble? First, it would provide for (and protect) a range of global public goods.[66] Public goods refers to anything that meets needs we all have in common, anything that entire societies have a right to and that no one individual or group is permitted to deny others: things like clean water, knowledge, good health. Even if some of us feel ourselves relatively secure and removed from the tyranny of hunger, we all require a "social shell" to survive, as the geographer Gerry Kearns puts it: and public goods do much of the work of providing for this social shell.[67] We are all reliant on such basic things as good soil and effective irrigation, functioning roads, and places to turn when we need help. We are dependent for our existence on "mutual aid", as the nineteenth-century Russian geographer Peter Kropotkin termed it. And GPI would be a far better way of addressing, for example, current food security needs than rich countries' trend towards land-grabbing and the corporatisation of the food-supply system.

In an increasingly globalised world, something is needed that can extend a public form of protection to the natural resources and the ecosystems that we all share, as well as to more intangible things such as the accumulated knowledge of public-oriented research (knowledge about life-saving pharmaceuticals, say). All of these things may be given a price and turned into tradable commodities. Many of them already are, such as the carbon that is traded between variously polluting nations. But public goods—even oil—cannot be safeguarded except by public bodies acting in the common interest. Bolivia's president Evo Morales recently, if unwittingly, raised the possibility of this when he told the world that Bolivia would desist from its plan to develop shale mining in the environmentally irreplaceable Amazon, if wealthier countries were prepared to incentivise Bolivia financially not to take advantage of that natural resource.

If we do not pre-emptively avoid the need for such blackmail, by ensuring that countries are all committed to an agreed-upon set of principles, we will see a lot more of this sort of thing in the coming century. Our present preoccupation with things like protectionist exchange rates and tariffs will by then come to be seen as matters of precious little importance in comparison, even to the political right. So perhaps it is now the left's turn to proclaim that there is no alternative. For while a powerful business lobby or private philanthropist may be able to stump up the cash needed to protect a sizeable area of rainforest (and even ask that Nestlé boats desist from anchoring there), in a volatile world there is no guarantee that corporations and non-governmental organisations will keep doing what they presently do or stick to their earlier promises. Democratic accountability really is our only hope.

The second benefit—of special relevance when it comes to inequalities in global wealth—is a distributional one. The University of Chicago economist Robert Lucas Jr. once famously said that questions of distribution are the biggest threat to "sound economics".[68] Given what Lucas and his free-market colleagues took for sound economics, that is precisely what makes it so important we learn to ask them again today. An international system of global public investment would give us a practical environment in which to do this. It would also, over time, help smooth out existing inequalities in wealth, in a way that is fair to the wealthy—not, for example, Soviet-style expropriation—*and* fair to everybody else.

It is essential that such transfers of wealth be reliable and predictable in the short term (so it is possible to plan around them) but equally sustainable over the long term. Our current model of aid cannot provide any of this: it has come to prioritise short-term funding and one-off transfers over longer-term investment. It is far too reliant on voluntary contributions. Similarly, foreign direct investment is too fickle and prone to market fluctuations: by its nature it tends to be pro-cyclical rather than counter-cyclical, making booms bigger and recessions drag out for longer. Between the two current dead ends of unregulated foreign direct investment and overly conditional official development assistance, therefore, global public investment is a fairer, more sustainable bet.

By providing a means of long-term, sustained investment carried out in multiple directions (overcoming the traditional problems of bilateral aid and conditionality between unequal "partners"), GPI would also be an effective down payment on the sort of global-scale trust that is required if we are to realistically tackle more complex problems—such as climate change. The more we partake in and disburse specified interpersonal (not personalised) commitments to other people, the more we will come to understand the nature of our common existence, no matter if our respect may continue to lag behind. This is how moral revolutions happen, after all: not with a bang but a barter. But it must all be enacted at sufficient scale if any of this is to be realised.

Third, GPI has the potential to ease increasing tensions over access to the resource endowments of poor countries and the geopolitical uses to which the territory of these nations is likely to be put. As the global food crisis of 2007 and the looming threat of water wars in the near future both make clear, absent clear rules of procedure as to how we organise the distribution of vital resources globally, and the capital required to exploit them, and the coming century looks likely to be one in which the recent decline in secular interstate wars and violence will be reversed. GPI would provide a much-needed preventive forum for managing the pressures that lead to conflict before war erupts. We have never found it hard to enter into treaties over contested territories in the past, the better to stave off conflict in the future; the problem is that we simply do not think to be as bold internationally, outside of military and economic bottom lines.

Fourth, our own belief in the value of society itself stands to benefit from a form of global public investment of the sort that I am proposing. We are not yet at the stage where it is even sensible to talk of a "global public" per se. Simply asserting that people become more interested in distant others than they are, absent a theory of why that should be, will always fall short of reality. Yet GPI would actually provide a way of getting us to the point at which we *might* begin to think like this. It is more important, under any definition of democracy, that we give each other our respect and our interest than our charity and our personal care. And by partaking, as relative equals, within a neutral system of taxation and distribution, we demonstrate such respect without the need to do or say anything at all. Our participation alone is enough. But we will learn this lesson only by the doing of it.

By doing away, at the same time, with the alms and moral guilt-tripping we are usually requested to bear, we might also come to see that not all the money raised would need to be spent "abroad". Some of it might finance research and development domestically, benefiting strategic sectors and industries—the UK pharmaceutical industry, for instance—in ways that better equip us for the coming century, or even benefiting more marginal and poorer regions domestically. In the GPI model, then, London's overheated economy in the south of Britain (a boon, for the most part, to only some of its inhabitants and those foreigners who park their investments there, the better to gain from the near 10% rate of return on property) could help divest resources to the north and bolster a more stable British economy in the process.[69]

Such arguments ought, then, to square off against fears that this is a game with but vanishingly small returns for wealthier and more powerful nations. Rather, there is every reason to believe that evening out imbalances in wealth between nations is exactly what will also make our wider world more productive again, and indeed more secure. It will certainly make our planet a more sustainable one, and it may even discourage large-scale international migration, particularly of the irregular sort: something in which the parties of the right can find something to celebrate too. After all, GDP had to be "made" a national obsession, as Harvard's Richard Parker points out: the same could be done with redistribution.[70] We aren't going to out-produce China, but shift the metric to some variant of GDP plus Gini and we might just find a means of showing that we out-redistribute them.

None of this is solely about economics at the end of the day. It is becoming increasingly clear, rather, that we need a more overtly political framework for addressing the long-standing structural inequalities of the international system. In times of crisis, we must of course prioritise the use and allocation of resources. But the point is that we need a more democratic means of doing this than at present. One way of achieving this would be to look forwards, not back, and in keeping with the social democratic ethos, to work on the terms of pre-distribution in addition to compensatory redistribution.

A DEMOCRATIC INTERNATIONALISM

If we want a world less tainted by the rot of inequality and unfairness, then we need, finally, to find concrete ways of including poorer citizens of this world in the decisions taken in relation to them, rather than simply waving abstract notions of freedom and equality about in the belief that doing so will encourage them to trust us. Institutional questions of democracy must come before any normative discussion of equality, that is to say, and since more and more problems can be solved only at the global scale, it is imperative that such questions be addressed internationally. It might once have been a valid critique that there was no necessary reason for national citizens to aspire to even a weak form of global politics, or to consider questions of inclusion, ownership and control beyond the borders of the nation-state, but there is no longer any alternative: we have put ourselves in a position today where some degree of international political coordination is essential. The sooner we recognise this, the better.

But what is democracy when it isn't the democracy of elective national states? It is certainly not the "democracy" that for half a century has been spread abroad in the name of peace and free enterprise. The history of democratic thought provides us, rather, with a spectrum of possible models we can turn to. For the most part, there are either federations (top-down models of the sort that bind the fifty states of America and the twenty-nine states of the Republic of India) or more loosely bound confederations (of which the European Union is perhaps the closest real-life example). But what is needed, today, as the thoughtful work of Daniele Archibugi, among others, has mapped out, is something in between the two.*

To be sure, this does not need to involve a fully cosmopolitan global order that lays claim to some singular global identity (whatever that might be). But we do need a global *institutional* system that builds on existing public and international law to better mediate between different political and social groups and across different economic and geographical levels, from the local to the municipal to nation-state and

*Daniele Archibugi, *The Global Commonwealth of Citizens: Toward Cosmopolitan Democracy* (Princeton, NJ: Princeton University Press, 2008).

beyond to the interstate. This is not about simply sticking local affairs together with global ones either; it is about recognising that the two already are implicit in each other. And the implication, as the political philosopher Nancy Fraser points out, is that democratic global institutions must marry civil society determination (or populism: the movements deciding the "what") with their own legal-political constitutionalism (to provide legally binding and institutionally effective resolutions on that basis).[71]

The trick to achieving this is to proceed backwards and step-by-step. First, we must build democracy back into such institutions as we currently have; we may then be in a position to build a more elaborate institutional architecture to address the underlying issues. At present we are heading in the opposite direction. For two decades the United Nations has preoccupied itself and the world with "peace keeping"; its earlier efforts to build the institutional conditions for peace have fallen quietly off the radar. More recently, since 2005, and the establishment under Kofi Annan, of the UN Democracy Fund—largely bankrolled by India and the United States—the United Nations has committed itself to Western-style "democracy promotion". But democracy should be more than a glorified talking shop of the more powerful and privileged nations. An effective second UN chamber—a parliamentary assembly—is sorely needed, and those of us with a voice in more powerful states should be using it to lobby for this. The idea is not so far-fetched—it has been the subject of concerted thinking for many years and has the support of at least one former UN secretary-general, Boutros Boutros-Ghali. More important, the idea has been endorsed by the regional parliaments of Europe, Latin America, and Africa and by more than seven hundred members of Parliament around the world.[72]

Such a parliamentary body would give the world's citizens a more direct say in the single most important international body that exists—the one body from within which a democratic focal point could be established and from which democratic oversight of other major international organisations (the World Trade Organisation and the World Bank included) could actually be exercised. A parliamentary body is certainly more appealing than the likely alternative: emerging powers, like Brazil and India, angling to ensure that they are the next in line to succeed the old structures of power (through demanding, and getting, permanent seats on the UN Security Council, for example). It would also be more effective.

Opening up the United Nations more directly to the people who are themselves affected by its actions (and today more than they ever have been) is one way to democratise it—or haul it into the twenty-first century, we might say. And in many ways this is nothing more than what the New International Economic Order was calling for. Another is to open up the United Nations to other significant non-nation-state entities that are the subject of its resolutions and directives. For example, more than 80% of the conflicts on the UN Security Council's agenda involve non-state actors—be these militias or governments in exile. As some have suggested, these actors should be given a "universal right of address", which would also require that they justify themselves to a wider public than those who are forced, by virtue of geography, to listen to them.

The potential here is not just for a more democratic way of resolving hostilities. Labour unions, channelled via the International Labour Organisation, could also be usefully included in this manner. Labour used to be the very heart of internationalism; perhaps because the world's workers are less and less "our" workers, it has slipped off the agenda in recent decades. But the future of our workers *and* their workers has scarcely ever been more closely entwined, and there is much to be achieved by putting work back front and centre on the international political agenda.

International politics is not, of course, limited to the United Nations and its affiliated agencies. A de facto international politics is orchestrated from *within* the core institutions of the Washington Consensus: the World Bank, the IMF, and the WTO. The WTO in particular has the power to shape the wealth of entire nations by determining the trade rules by which states are to be governed. At present its demands work almost entirely counter to the forms of social policy and market regulation we have discussed here. And they consecrate as international charter a fundamentally unfair economic system.

There can be no doubt that this needs changing. But a more broadly based perception of the problem is required first of all. Supporters of free markets—and the WTO is nothing if not their Cardinal Richelieu—like to claim that, at the end of the day, all that matters in policy are the incentives. Get the incentives right, and people will act appropriately. That is why regulation is superfluous, if not dangerous, they say—even as "incentives" lead nations to lower their minimum wage. Do these people not see that regulation is itself an incentive for us to act in the interests of people other than ourselves? Do they not see that self-interest does not itself preclude acting in the wider public interest?

For all that the WTO's positions are easy to critique, they are incredibly hard to shift, because the countries that are most affected are the ones that have least say in the system. In the past, if people really wanted to change something, it was enough for them to march on their capital city. Today, after storming the barricades in Tegucigalpa or Macau, those marchers would likely find that they had to put a phone call through to the appropriate representative of the WTO or the World Bank. They would be forced to rummage through the filing cabinets to pull out the small print of exactly which rights had been given away in the latest free-trade agreement with Europe.

The problem is that, for many issues, it is simply no longer clear just to whom exactly we should turn to complain when things go wrong, or whether those to whom we complain would even be permitted or capable of doing anything about it anyway. And the fact that this is not clear says everything about the extent to which they were determined democratically. We are told that international politics is a pipe dream. But in fact international politics is alive and well—it is just not a very *democratic* international politics at present. And ironically, the people telling us that there can be no international politics are the ones who are doing most to actually shape things. So international politics exists; at present we merely act as if it didn't.

The problem of course is that people cannot alter the great ships of state and interstate institutions alone: they must come together somehow. But the most effective institutions we have for educating, encouraging, and coordinating such desires—unions and mass organisations—are the very institutions that are also on the decline. In the United States, for example, labour union membership declined by 43% from 1950 to 2000.[73] In Europe it has fared only a little better. The figures for the number of people who turn out on strike follow a similar trajectory. Absent such institutional focal points around which to converge, to bring pressure to bear on entrenched interests, and through which we can ultimately settle upon common policies, we lack a mechanism for even identifying the rights that we would have protected in the first place.

It is little wonder, then, that in place of the demands for universal policies that mass organisations once made, we have switched instead to a focus on elective social movements and civil society groupings. There is much that social movements can achieve: London Citizens and Citizens United in the United States are two important cases in point, as are *buen vivir* in Latin America and *beni comuni* in Italy. And civil society has an even greater, untapped potential still. But single-issue initiatives are vulnerable to being out-manoeuvred by governments. In Peru, social movements mobilised against extractive industries but were easily out-played by a government that deliberately framed the national "anti-poverty" discourse in terms of export-led (which is to say, mining-led) growth.[74] In Argentina, communities mobilised on ethnic grounds, but their achievements were then used by the government to wriggle out of *universal* policy commitments.[75] International movements which can transmit expertise are needed to combat this. The World Social Forum and the International Transport Workers' Federation each offer ideas as to how. But these are not what most people think of when they think of international institutions.

The original social organisations within the United Nations are equally important and equally under-valued today: the UNDP, UNCTAD, and the ILO chief among them. It is not their fault that they have lost the regard that they once had. But it is to all our cost, since in their absence, too much international politics has come to take place behind closed doors. Worse than this, too much is actually privately determined by the super PACs and the business lobbies. When we remove the United Nations, which deals with policy but which today prefers to do so on an ad hoc, crisis-response basis, and the regional political or military organisations (such as NATO), none of which deals with social policy-making in any significant way, the majority of what is left, which is to say the organisations that actually *do* things and have the power to make other people do things, are business concerns: the institutions of the Washington Consensus (the IMF, the World Bank, and the WTO), the World Economic Forum, the African Economic Forum, NAFTA, APEC, a range of bilateral and regional free-trade agreements, and elective groupings such as Business for Peace.

Against these Roman legions of the business world, we need to find ways of building democracy back into the existing international public architecture: keeping those bits which are useful and can be reformed, and doing away with those that cannot. This is not something these international organisations are themselves necessarily against: there are strands within the IMF that are working on increasingly progressive approaches to international finance, and it was the chief economist of the World Bank, after all, who stated back in 1996 that "reducing inequality not only benefits the poor immediately but will benefit all through higher growth."[76] These institutions should be held, as a first step, to these voices of their better angels. But reform is ultimately needed.

Reform of the United Nations, via the introduction of a more democratic chamber and renewed appreciation of the role of its more progressive wings—the UN Development Programme of old, the ILO of today, as well as the UN Industrial Development Organisation and UNICEF—is one way to go. Another way is to build out the regional level of international policy-making. Policy powers must be taken back from regional economic agreements and given over to organisations like the Bolivarian Alliance for the Peoples of Our America (known as ALBA), Mercosur, and the Association of South-East Asian Nations. Social policy protections must be ensured at the national and regional level too (something that is almost entirely overlooked in the MDGs' and SDGs' focus on reducing the poverty of *individuals*), and that in turn means making it harder for wealthier nations to co-opt the WTO's dispute and enforcement mechanisms.

The experience of the European Union is instructive here, albeit something of a mixed example for would-be cosmopolitan social democrats. The original European Community was never intended to engage in social policy-making, but it ended up having to when its own promotion of the free flow of people across borders played out in ever more socially disruptive ways. And to its credit—for all that it is out of fashion to compliment the European Union these days—it actually did. From 1988, for example, the budget for the European Social Fund increased year on year with money drawn down from the Common Agricultural Policy and from the EU entry of wealthier countries like Sweden, whose own history we are by now familiar with.[77]

Regionalising democratic decision-making is but part of the solution, however. Ultimately, some sort of independent global public institutional infrastructure is needed, no less than nations needed roads and railways in the past. Simply put, we just will not get much further without it. Unions and social movements, and indeed NGOs, may all have a role to play here. But the lead should be taken by UN-established bodies, which are often the best positioned—and the most competent—to deal with the specific challenges that await: be it labour, migration, climate, or health.

Again, GPI is instructive here, because it could itself become a vehicle for developing an integrated, cross-national institutional infrastructure. In many cases, as we have seen, poor countries' governments simply do not have the means yet to develop sufficient tax systems, or even a sufficient tax base: only 5% of the Indian population is wealthy enough to pay taxes. So it is clear that these governments are likely to need some help: not least given that the other alternative, a form of Keynesian deficit financing, is effectively closed to them in today's economic climate. This changes things a little. In the past nations incurred a debt to build a railway with the expectation that in the future they would reap a greater series of benefits. Today, when it comes to the public realm, wealthier nations must empower poorer nations to help us all reap those future benefits.

Finally, there is the question of the relationship between democracy and the law. From the scramble for Africa to Europe's great era of constitution-making in the 1920s, the law has always been the silent party behind the founding of new social orders. In the post–Cold War era international jurisprudence has again experienced something of a boom. But that boom has to date adopted the normative tenets of liberal democracy as its moral compass. We are familiar by now with some of the effects.

The constitutive power of the law can be made more democratically accountable than this. At present it is corralled into a (somewhat overplayed) struggle between the might of "constitutional" orders, which are invariably associated with particular places (Brussels usually), and individual human rights (which as we have also seen are associated with a universal freedom of choice untainted by the grit of actual political commitment). It is probably fair to say that the benefits of constituting orders have been subsumed by unjustified fears of their effects while the pitfalls of human rights have been buried beneath the celebratory rhetoric with which they are usually deployed. It is almost entirely overlooked for the most part, for example, that human rights discourses have themselves frequently acted as a barrier to the extension of politics internationally.[78]

There are further problems here too. Americans on the whole are more sceptical of international law (as the recent history of US unilateralism reminds us); Europeans, in contrast, tend to be more in support of it (it is the *European* Court of Justice after all, and the International Criminal Court is largely the child of Europeans).[79] Part of the reason that human rights have become such a dominant and successful political language since the 1970s is precisely the fact that they are able to square this circle: giving Europeans and Americans alike a little of what they both want. But this circle cannot go on being squared if we want to start talking about constitutional forms of politics. The problem goes deeper than just the limits of human rights.

The problem, at its heart, is this. The constitutional reach of international law has to date been driven primarily by (European) *experts*, not by a process of democratic deliberation.[80] And this matters, because the well-known American reluctance to support a fuller role for international law—it is one of the few states that has not ratified the Rome Statutes, which give the ICC its authority to prosecute, for example—stems in no small part from the belief that the sanctity of democracy outranks, and needs protecting from, the tentacular constitutional reach of international law.[81] On the surface of it this is a reasonable position: it was in the European Court of Justice, after all, that British

Chancellor George Osborne first tried to challenge the proposed European Financial Transactions Tax.

But it is also revealing of the fact that it is a distinctly conservative vision of democracy that holds most strongly today in America. And it reminds us why a conservative form of democracy will not help us to solve the problems of this world: we *do* need to take the risk of empowering international law, and that means that we must also take the risk of allowing national democracy to be rebuilt in line with this via democratic constitutional processes. That means not seeking to prevent international law from constraining what states can do (the US position) and not seeking to override national democracy (as does Europe at times) but getting the twain to meet. The idea that a country can have either a constitutionalist approach to making the rules or a democratic approach is a false choice, and we should reject it as such. The experience of the former Soviet states transitioning from communism to democracy in 1989–1991 and South Africa's democratic transition in 1994 both show that there are ways that the making of new constitutionalist orders can be undertaken in a democratic way *before* a democratic polity as such has itself been constituted.[82]

International law is important too for its capacity of oversight and redress as much as for its constitutional ability to "create" new orders. Perhaps above all, the current laws pertaining to intellectual property rights need addressing, so that what ought properly to be enjoyed as public goods—certain medicines, for example—is not kept locked away behind monopoly rights for the benefit of the rich alone. It could be used to revive an updated form of anti-trust legislation, say—and to make firms and others accountable not only as producers of exchange value but as employers and shapers of use value too. It could be used to underpin anti-land-grabbing and anti-social dumping rules.

We know, then, the sort of thing that is required, which involves a *political* response to the challenge of global inequality. We need to re-create the vista of 1944 but unpick our ideas and the values that we have inherited from the 1970s in order to do so. The solution to the problem of uneven development and global inequality requires of us that we build new institutions and democratise existing ones, but not just in and for the poor world. The rich world too must be party to and bound by these changes, which is why international law is so important.

We need new "cow trades" to address the unfair terms of trade under which most of the world's people are forced to labour, and not least because Western citizens also lose out to the current corporate subsidies (it is we, after all, who are paying for our governments to reduce corporation taxes). We need publicly minded economists and lawyers to devise and put forward a just global intellectual property rights regime so that people who are sick in poor countries can afford the medicines they need, and so that poorer societies can develop their own domestic pharmaceutical sectors offering drugs at prices that Western consumers would no doubt prefer to pay as well.

We must be prepared, in short, to do the hard work of joining up the dots of what is, at present, a far too patchy, far too easily manipulated institutional framework that governs the lives of rich and poor around the world but that does not govern them alike. The global poor are marginalised within the current system differently than the wealthy world's poor are marginalised within it. But the end result is the same. Both would benefit from greater equality. Conservatives may say that progressive forms of taxation, which might underpin this, are a limitation upon some (rich) peoples' freedom. But the belief that underlies this is wrong, as Ronald Dworkin, among others, demonstrates. It is also the reason that equality has always ranked second to liberty in the modern world, and this, in turn, is something the rest of us lose out from.[83]

So we are justified in seeking greater equality. This politics or "strategy" of equality must be universalist in aspiration; morally, it cannot allow gains in one place to be the product of injustice elsewhere, but practically, too, it will find itself otherwise severely limited in addressing the injustices of a modern globalized economy and such problems as tax avoidance. It must involve ceilings on the power and wealth of the rich. Political power is a more difficult thing to dislodge; economic wealth no less so. But ways can be found if we are prepared to embrace a new vocabulary of change: for all that human rights are a powerful mobiliser of empathy, recognising their particular limitation here is essential. If this strategy of equality is to be meaningful, then it must be democratic and participatory. Since it is impossible to specify a single vision of the good life, this must be iteratively determined and able to change tack as society develops.

There is no single way of reaching this point, and since we are not aiming for a single point in any case, there is no utopia to peddle on its behalf. What is required, rather, is a willingness to act. Something needs to stand behind these changes in our current ways of wealth, in our attitudes to pre- and redistribution, and in our commitments to democracy and the law. At the end of the day, that something can only be us.

6

"THE SOFT POWER OF HUMANITY"

How can you expect a man who is warm to understand a man who's cold?
—Alexander Solzhenitsyn, *One Day in the Life of Ivan Denisovich*

"Whoever invokes humanity wants to cheat", one of the paragons of political realism, Carl Schmitt, once wrote. Yet Adam Smith was not cheating in *The Theory of Moral Sentiments* when he turned to what he called the "soft power of humanity".[1] For Smith—hailed by many as the father of free-market, laissez-faire economics—to be talking of the *limits* to market rationality and the "authority of conscience" strongly suggests, in fact, that cheating was the furthest thing from his mind.[2]

He was, however, being rhetorical. For Smith would have words with both the "whining and melancholy moralists . . . perpetually reproaching us with our happiness, while so many of our brethren are in misery", and with their counterparts, those who, as he put it, only ever think to ask, "To what purpose should we trouble ourselves about the world in the moon?" His words could hardly be more apt today. For Smith, what should really guide our relationship to the strangers who—with us—make up humanity, are not the faculties of sentiment and despair, and much less the famous invisible hand, but instead "reason, principle, conscience", as we may impose these upon ourselves in dialogue with others. In matters of political morality, Adam Smith was a social democrat.

DO WE NEED A GLOBAL PUBLIC?

What Adam Smith saw more clearly than nearly anyone else in his time was that the social and economic changes wrought by capitalism had changed the world so profoundly that political thought and practice had fallen behind the times. Success in trade was so important, for example, that it had become a matter of the political and military survival of nations.[3] There was no point replaying the same old debates over and over. Quite simply, the world had moved on.

The world has moved on again. The current moment presents us with an escalating series of seemingly intractable problems: environmental degradation, persistent poverty, a crisis of faith in the "public" domain, a loss of trust at the interpersonal level, everywhere the familiar *nosismo* of us versus them.[4] Our economies, our technology, our own patterns of thought, all have rushed ahead of us. Our politics is struggling to keep up.

So what are we to do? We might turn to the history of capitalism and the long-run dynamics of wealth accumulation to confirm the inevitability of all this. Or we might devise ideal-type societies in accordance with whichever model of social justice we happen to prefer and then use those to highlight the deficiencies of our own. These all provide useful analytical tools and a greater measure of clarity as to what is at stake in an increasingly unequal and fractured world. But they do not themselves disclose, much less diagnose, the deeper challenge before us. They cannot do so, because the very idea of political society these analyses operate with no longer holds.

Smith's own response, in his time, was to shake the very premise of "public good and civic virtue", upon which the institutions of states, markets, and society had been built in the mercantilist era, and replace it instead with a new liberal emphasis on popular sovereignty and individual freedom. In some respects, the historical struggle amongst Marxists, liberals, and social democrats that has raged across the two centuries since has itself merely been about negotiating, in different ways, the terms of that liberal settlement. But the task today is different: it is about trying to rethink, and reinvent, the terms of the prior settlement. If we wish to address what really ails us today, then we are called upon to reinvent the political forms of modern society for our own global age.[5] Yet we are challenged even to *imagine* what that might involve.

The primary aim of this book has been to try to address this hurdle of the imagination, and to suggest that the political reinvention we require today, in *this* world, is most likely to succeed if, first of all, it takes its cue from the sort of democratic compromise and institutional innovation that has been most convincingly demonstrated to date in the historical experience of the social democratic *project*, and second, if it finds ways to successfully internationalise those lessons. The book's further claim, however, is that since rich and poor alike stand to gain from such an undertaking, the incentive for this reinvention already exists.

Let me be clear about what I am not calling for. It is neither possible nor desirable to expect people to snap into line as some happily constituted "global public", much less to do so in response to any one person's vision of what that public should be. What we need in place of this is a rationally defensible account, first of all, of what our individual rights and collective obligations are relative to those distant others with whom we may hold meaningful—if not always visible—attachments. Second, we need a clear sense of the sorts of institutions that will be required to manage these relations. These are both fundamentally political questions and they each require that we have in mind some kind of model of the political community we are after.

The most well rehearsed and comprehensive of these models is cosmopolitanism: the idea that one's commitment to the idea of humanity itself will ultimately be able to compete with more narrowly imagined and more concrete claims of nation, race, or class. In fact, the state remains quite central to many versions of cosmopolitanism, not because it represents an ideal political form, but because it is often just about the only tool capable of challenging the power of the market. A common critique of cosmopolitanism, however, is that it is inherently elitist, if not wishy-washy, and obscure. There is certainly some truth in this. But there is greater truth in the fact that numerous forms of cosmopolitan attachment already exist between people and groups the world over, even if we don't immediately think of them in this way. A more serious problem is that it offers no real answer to the questions "Who am I?" and "What do I want?"—questions that matter to young people most of all. It therefore lacks political edge.

But if the idea of a global public as the cosmopolitans envisage is too diffuse to catch this or that political wind needed for it to carry throughout the world, it is equally clear that the convergence of self-interest underpinning state-based "coalitions of the willing", the global "compacts" of international businesses wearing their corporate social responsibility hats, mega-philanthropists like Bill Gates, and even such global "commitments" as the new Sustainable Development Goals is too narrow and too undemocratic. The same applies to bottom-up special-interest communities that underpin a good many social movements—be they patient groups or social justice campaigners—and to the current vogue for philanthropy. These agents' attachments to one another and to their issue may well be real. But the community they build is an elective one, and in that sense, it is no different to online communities of people "liking" and "friending" one another because what they do mirrors back what they would themselves like to be. Their community may be public but that does not alone make it democratic.

The world is, or at least should be, more than just an open frontier of opportunity upon which to project the self. There is today a desire to "engage" with the world (it used to be enough to want to "see" it). But engagement with the world soon gets tossed onto the mental CV we are all encouraged to keep. For all that Tahrir Square has been hailed as the Bastille of the Facebook generation, the more pervasive effect of social media has been a form of anti-politics, for example: it becomes possible to watch everything up close, to "follow", even get to take part in real-life political theatre. But it is much harder to actually "build" things on social media.

The real virtue of social media platforms is that they demonstrate the connections that already exist between us. The global response to the deaths in the United States of Michael Brown and Eric Garner in 2014, for example, lit up Twitter accounts the world over in a glowing rage of solidarist hashtags: #BlackLivesMatter, #ICantBreathe, and #HandsUpDontShoot. But this can also be their great weakness. Most of the time, when we follow and like, we create a world, often quite an international world, around us. But we experience, via a narrow portalling of our social interactions, only a very thin strip of that world. Simply put, most of the more than 1 billion Facebook users are not following the other 5 billion people on this planet. They are following people like themselves or people they would rather be.

The pendulum of "engagement", then, all too easily swings the other way. Rather than everyone taking part in the decisions that really matter, we are all encouraged to leave it to those who want to and can. The esteem with which Bill Gates is held for his philanthropic work is indicative of this. Yet philanthropic giving is not a democratic form of charity, as anyone familiar with the history of philanthropy since John D. Rockefeller would attest.[6] As the Norwegian philosopher Jon Elster points out, the only true giving is any case anonymous giving. And things like global health projects should be a matter for much wider public debate than Gates is known for encouraging.

Rather like Louis Vuitton's "global values", the current wave of megaphilanthropy, from Donald Trump (whose commitment to the causes he professes is apparent from the fact that he is not even the main benefactor to his own foundation) to the Giving Pledge overseen by George Soros and Bill Gates, is as much about brand as it is about beneficence.[7] As the public health campaigner David McCoy points out, the word "philanthropy" literally means love (*philia*) of human kind (*anthropos*), and yet more than 45% of the $500 billion held in 2008 by US philanthropists was money held "in foundation"—where it sits peacefully beyond the reach of the Internal Revenue Service.[8] Virtue is of course a dangerous thing to legislate for, but we are no better served when virtue is simply priced out of reach for all but a few of us.

In truth, neither the universal abstractions of cosmopolitanism nor the individualising assumptions of philanthropy will solve the collective action problems that confront us in thinking about how best to tackle the unequal state of our present political order. What is required, instead, is the means to sustain a weak idea of global public-mindedness that falls somewhere between the two positions: a sense of reciprocal coexistence that is neither as grand as a universal collectivity of "all" the people in the world, as the Nobel Prize–winning economist Amartya Sen has suggested, nor as specific as nation-states or self-asserting interest groups.[9] The former is impossibly broad anyway, and the latter no protection against the internationalisation of lobby interests. One thing we certainly do *not* need, let us be clear, is a global version of the National Rifle Association.*

*This is not an idle point; the Indian business lobby, for example, is becoming steadily more Americanised.

As present we are twice removed from this: the international decision-making that does take place and that is carried out with a wider, non-national public in mind is invariably executed by the least democratically accountable institutions (the UN Security Council, the G8, the International Centre for Settlement of Investment Disputes, and so on) while organisations that are more publicly accountable lack the power to enforce their decisions (the somewhat neutered UN General Assembly). This needs addressing, as we have seen. But above all we need to find a way of enabling collective action when the vernacular of public spirit seems to have so little purchase any more.

POLITICAL RESPONSIBILITIES

"For mutual understanding," wrote the Hungarian author George Konrád at the end of the Cold War, "tolerance is not enough; one also needs *complicity*. . . . We can share our experience only by living it together." Konrád's words, translated to our present situation, stand as a warning against the passive "tolerance" we seem content to offer the world and its troubles: a system of safari aid donations and "fact-finding" missions that dares not open up any real democratic debate about what is actually needed and what we, in particular, may contribute towards this. What is required in place of the present status quo is for us to start talking about our *political responsibilities* rather than our *moral obligations*. Until we do so, the inequality divides that separate humanity today will remain as with us, as ugly, and as firmly set in concrete as was the Berlin Wall that divided Konrád's world.

But what exactly are political responsibilities? As the philosopher Iris Marion Young wrote, political responsibilities are obligations we incur by virtue of our structural position in society.[10] This can be as consumers whose demand for cheap clothes is the foundation of an unfair commodity chain, or as men who, by virtue of their gender, earn more at work than women (to speak only of income inequality when just 20% of all parliamentary seats worldwide are held by women is to miss a very significant part of what it means to talk about inequality). It can be as Western couples who sit at the top of a care chain passing their own domestic economy response to the declining welfare provision in the West onto that of their Filipina maid and her family, who in turn must manage without her care the other side of the world.[11] We have seen many examples of these sorts of responsibilities in the making in this book. But in almost every case we can meet these political responsibilities by making the sorts of institutional commitments towards one another that I sketched out in the previous chapter.

The usual rebuttal to arguments like this is to dismiss them on the grounds that they cast the net too wide: holding innocent people guilty on behalf of weak and unspecifiable problems of "many hands". But there is actually no need to insist that we are all equally enmeshed in everything. There are *degrees* of political responsibility, and different obligations follow from those according to how any one individual is positioned with respect to them. As a meat eater, for example, I have greater responsibility to address inequalities in animal-derived protein consumption than does my vegan neighbour.

Similarly, as a "bit-player" rentier, to use Andrew Sayer's phrase (I have a pension invested in financial markets and I have benefitted from those increases in London housing prices), I have a greater obligation than somebody living in council housing to demand regulation of global finance and a proper system of property taxes in the United Kingdom.[12] As the prime users of tax havens, Europeans have a particular obligation to desist in using them. Guilt, to the extent that it should be a part of this, is not personal guilt (as we commonly understand it) but rather what the political scientist Ruth Marcus calls "existential guilt": a way of acknowledging one's material connection to a particular form of injustice. It is not about forgiving the debts of others so much as acknowledging our own debts and seeing the process of paying those debts back as one that we might stand to learn from.[13]

Such formulations are little known to economists. But they are well known to historians, philosophers, and political scientists, who have often raised them in relation to past political atrocities: be it the Nazi genocide, or colonialism, or the aftermath of transitions to democracy from authoritarian regimes (as in post-Pinochet Chile). The guilt or otherwise of the German population under Nazi rule, for example, was discussed in a series of letters exchanged between Karl Jaspers and Hannah Arendt after the war. They both differed in important ways in their thinking about "the German problem", but they agreed that political responsibility had to be separated from personal guilt, and that what mattered most was liability (which individuals could hold whether they were aware of it or not). There was a need to counter such liability, they agreed, by forms of civic responsibility, up to and including the modification of one's own understanding and conduct in light of what could no longer be denied. It was not the abnormality of the crimes that was their focus, but the very normality of peoples' relationship to them.[14]

The experience of Chile in the years after Pinochet's rule relates these questions of political responsibility to matters of political economy even more directly. In the Chilean case, an active recognition of the impossibility of assigning personal guilt (just who *was* part of the former system?) led the bargaining process (between former elites and the new representatives of the Chilean opposition, the Concertación) to settle on a collective determination of guilt, based on class interests and resulting in—of all things—a relatively progressive taxation system as the most appropriate form of reconciliation.

That forward-looking desire to hardwire the recompense for past wrongs into the future good of the nation has been central in helping reduce inequalities in Chile (look again at the figures on pay dispersion in Chilean manufacturing) and in underpinning Chilean development ever since. In a situation when so few individuals could really be said to be entirely free of the taint of the former regime (because almost everyone, after all, had to accommodate it in some way or other—they were all, in that sense, complicit), this was a pragmatic and a positive step. The idea was not to blame any one individual or group but to invest the energy of the transition process into removing some of the underlying problems of an unequal society: the very problems, in fact, which had nourished the earlier political violence in the first place.

It is political compromises of this sort that are required today if we are to address in any meaningful way the structural injustices from which global inequality stems. If the neoliberal doctrine that has most strongly shaped our political age ultimately operates on the basis of a model of distributed responsibility (a problem you're familiar with if a telephone operator has ever told you that another company now runs the service you've called to complain about), then the solution must be its opposite: to *share* responsibility. "Try as we might to live well," the philosopher Christopher Kutz points out, the great problem with our modern age is that "we find ourselves [unavoidably] connected to harms and wrongs."[15] In short, if nobody takes responsibility, we will all end up paying the highest price; but if we all take responsibility, the cost of our doing so will be shared. What is needed, in light of this, is not moral virtue but a political orientation to the idea of civil responsibility. We are not wrong to enjoy our privileges, in short, but we must take responsibility for their systematic consequences for others.

What the above examples show, in the vernacular of moral philosophy, is in many ways all that history has taught us time and again and mainstream economics now needs to learn: adopting more socially just behaviour does not require us to change who we are but to create new habits on the basis of pragmatic consensus. Moral heroics alone did not bring an end to slavery or Jim Crow, and they have not still to institutionalised racism. It was not William Wilberforce or Rosa Parks who brought about the wider change in their societies. They each crystallised an issue; they became *vehicles* for change. But in both cases, shifts in wider public *conventions*, the agglomerated effect of the civic consideration of individuals, made practices that once seemed inevitable and normal—even though there were already rational, and strong, arguments against them—into practices that were no longer tolerated and practiced. It is a difference of degrees. But within those degrees lives are weighed in the balance.

So we do not need more arguments about exploitation, or even neglect—our two primary modes of critique, important though they are.[16] What we need is clarity, first and foremost, about where each one of us touches some part of the wider structural injustices that blight our world, and then a willingness to act differently in relation to those injustices. Both point us again to the fact that it is not charity that begins at home so much as responsibility. If we are all, to differing degrees, connected unavoidably to structures that exploit some while privileging others, then we all have some ability to influence those structures in our own here and now. Recognising this does not limit our freedoms; it offers, rather, the opportunity to exercise the full faculty of such freedom as we actually have. *This* was what Karl Polanyi was getting at. He saw that we are most free when we agree to limit ourselves. In many ways we have arrived back to this point today. The taming of global markets and the reinvention of the state will both be possible only when we change our own disposition first of all.

THE GLOBAL SOCIAL QUESTION

As he looked upon England's territory in the nineteenth century, Benjamin Disraeli was moved to remark, "The social question is today only a zephyr which rustles the leaves, . . . [but] it will soon become a hurricane."[17] He was referring to the same sense of unease that Adam Smith had put his finger upon: the unease of a modernising society, wrought by class divides and no longer sure of its fundamental values. Disraeli was right to be concerned, as it happened, for the likes of Vladimir Lenin were indeed waiting behind the tree, as J. Edgar Hoover might have put it, for the storm to arrive. The winds of change are that much stronger again today. But perhaps this time the storm need not be as we imagined, and lived, it during the twentieth century.

The social question, as understood in nineteenth-century Europe, was traditionally a national question: *die soziale Frage* in Germany, *la question sociale* in France. The social question took shape as the wealthy began to recognise and to relate to the plight of the poor and to ask themselves what, if anything, they were to do about it. Their concern was in part self-regarding, and it was also a distinctly partial preoccupation: outside of Europe the colonial authorities had more interest in extracting value from their charges still than in securing their capabilities. To the extent that those other countries themselves later asked their own version of the social question in the post-colonial era, their efforts came to a rather unceremonious end, as we have seen, around the 1970s.

The question of the relationship between wealth and want is equally pressing today, but it appears this time as a distinctly international problem. In that sense the growing problem of global inequality and the crisis of the Western welfare state are two sides of the same coin. There is little sense, however, that they are being recognised as such. The old tropes of means-testing and handouts for the poor are back with a vengeance in rich and poor countries alike, while the "benefits"

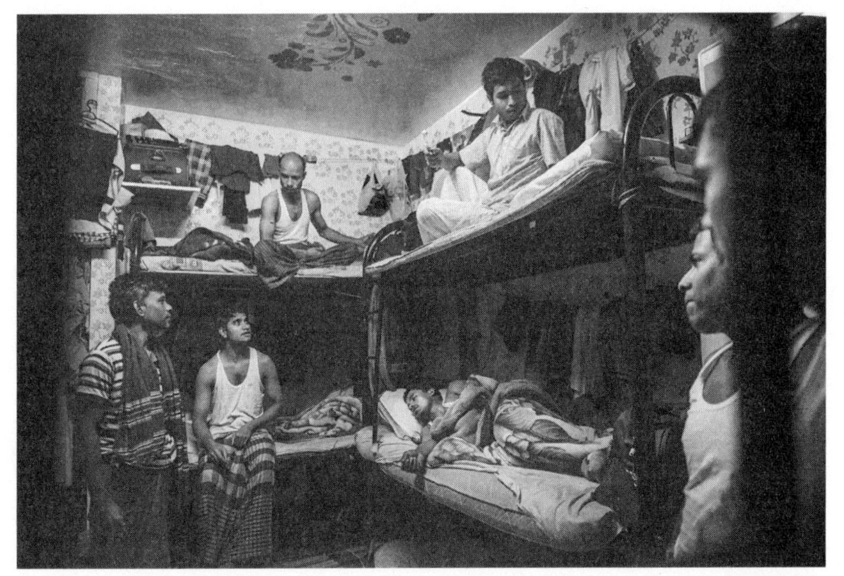

FIGURE 11. The Dickensian past in Abu Dhabi's present. Migrant workers from Bangladesh share a room at an apartment where they live with other workers in Abu Dhabi, United Arab Emirates, April 13, 2014.

of underpaid reserve armies of labour are being rediscovered in the construction booms in places like the United Arab Emirates. Unless we find some way to address this, then our governments and corporate leaders will remain tempted to keep playing off the poor of some places against the poor of others. And we will all keep losing out as a result.

But what does addressing it actually involve? As I have tried to show, to the extent that our economies and our societies with them have been globalised, our politics needs to follow suit. In the first half of the twentieth century, we have seen, W. E. B. Du Bois believed that the struggle for equality for blacks in America was of a piece with a wider anti-colonial agenda.[18] By supporting the poor of other, blacker lands, those discriminated against in the richest nation in the world would achieve a wider constituency of support at the international level, which might pressure their own government into granting them their full uninhibited rights.

Du Bois was right, for all that the international relations of the period made his position a hard undertaking. Christian Aid today makes a similar case for why we should fund civil society institutions like Tax Justice

Network in Africa. They are right too. But while such cross-national forms of solidarity would seem to be more easily put into place today, they are frequently also held back by the obstacle of contemporary public debate muddying our understanding of the unavoidable relationship among states, markets, and society. Removing this obstacle is central to task of the resolving what is, we can hardly deny, the *global* social question that confronts us today.

The immediate problem we encounter here, however, is that in almost all modern efforts to specify some desired notion of a better society, states and markets tend to be kept conceptually separate (at the same time as one or the other of them is presumed minimised in the final arrangement). Our present, much-polarised debate is certainly unhelpful here. But it is also itself a product of this wider problem. As we have seen, it leads us to ignore the fact that it is the nexus *between* states and companies that has historically been the most damaging (from the lobby system to government concessions and corruption of every sort). It is obvious that both states and markets have their deficiencies, and even more obvious that these are most quickly discovered when the two are left alone in each other's company, free of public oversight. But this is not a problem we can wish away; we simply have to solve it.

When Adam Smith was writing, there were good grounds to be suspicious of the state and to wish it cut back. It was a time when government was riddled with special privileges, when rules were little more than plutocratic bulwarks, when monopoly was rife because powerful individuals almost always had political (read: class) connections. But as he had the wit to recognise before many others did, there are also some things—usually necessary things—that no one *but* the state will pay for. There are such "public works", he said, "which, though they may be in the highest degree advantageous to a great society, are, however, of such a nature, that the profit could never repay the expense to any individual or small number of individuals".[19]

And yet it is also true that, in our world, markets have frequently become a more powerful means for the projection of power than states could ever be. The in-built failure of classical economic theory to recognise this—to ignore questions of power almost entirely in favour of a discussion about rational choice—is a major reason elites have long been able to get away with abusing that power. And yet if, paradoxically, we can be free of the state only through the state (through the protections a state agrees to submit to), then the same applies to the market (with respect to the regulations that must have universal reach if they are to be effective).

Just as we need more of the right sort of state, then—not kleptocratic states or states that like to borrow from the wealthy instead of taxing them, but states endowed to provide public services that work and with functioning democratic systems—so do we need more of the right sort of markets, not less markets: markets that favour Main Street over Wall Street, cooperatives over corporations, and where intervention is permitted on grounds of public need and not because some entities have become too big to fail.

There is in fact no necessary reason private actors, as a group, should choose to maximise profit to the exclusion of all considerations of social justice. Markets are broad and complex ecosystems: they are more than just a bell jar of Mandeville's bees. They are the conditions of existence for organisations like the Congress of South African Trade Unions (COSATU), which did so much to bring down apartheid in South Africa; and they provide us with a means of expressing our identity and our relations with other people outside of communitarian norms. But we need distance and space from them if we are to realise this: and that means ditching the idea of wealth as the self-evident end of all our means. It means accepting to become more political once again.

But if neither states nor markets should be allowed for long out of the sight of democratically enfranchised citizens, what makes us so sure that "the people" always act in the interests of one another? Anyone familiar with Henrik Ibsen's *An Enemy of the People* will know that the mob, even an educated, *Guardian*-reading mob, can be an unpleasant force in the world. Tyrannies, be they majorities on the march or minorities at plot behind velvet curtains, always need states to hold them in check and a market to give vent to their desires, preferably before those desires get bottled up as frustrations. As Onora O'Neill puts it, what we need is not more trust per se, but more trustworthiness.[20]

In all the talk about the power of civil societies we sometimes forget this—that the society we like to cast in the role of vanquisher of state or market is itself a product *of* states and markets. More than this, society needs both markets and states to realise itself. This is not a bad thing, however. It is true that social movements can be a source of democratic learning for citizens, and may well prove especially important at times of transition.[21] But it is also true that without states or markets, civil society has very little purchase on the world at all.

Nowhere is this clearer than in the fact that, for all civil society has been on the march in the post–2008 crisis era, for all the talk of hashtag activists, there has been on the whole too little plough-through from these movements to actual political change.[22] Why? Because until these movements found a political vehicle—be it Syriza or Podemos—the hard work of establishing what was actually possible could not begin. As Gøsta Esping-Andersen pointed out at the very start of our post–Cold War era (if only we had listened then), we all want protections, and we want them not just as individual citizens: businesses want protection from striking labour; farmers want subsidies to protect themselves against imports of other farmers' outputs; and nations seek protections, via tariffs, for core national industries.[23] And this means that, just as the post-Soviet states had to find democracy before democracy, we must now find equality before equality. We need, to put it succinctly, to change the "why".

AN AGENDA FOR CHANGE

We are returned, then, to the value of the social democratic insistence on constituency mobilisation and democratic negotiation between "equalitised" parties as a model for political change. The legitimacy of the case for change itself is not enough. We must now implement actual changes. Politically, the origins of inequality stem, as we have seen, from the capacity of the already privileged to lock into the system their advantages over others: but in a society organised according to the prioritisation of wealth as an end in itself we are all complicit in that system. It is true that some of us may be more privileged than others, but we are all sometimes the excluded and all sometimes the excluders in this game, and so we must find ways to lock fairness back in to our national and our international politics, together.

We must recognise, in doing this, that markets are always political, whatever some may argue, and that societies of free and private citizens are made possible only by public laws and the democratically determined means of upholding them. We must acknowledge that while we are all different, and no one has quite the same needs, we progress our individual needs best when we accept to work with others. At the same time, there is no universal morality that binds us to the categories of "us" and "them": our politics must be an engaged one and not just a confrontational one if it is to succeed. The most equal and the happiest societies today are, as I have painted things in this book, but as any number of quality-of-life indices will also reveal, those that have come to accept all this. This ought not to be so surprising. Since states, markets, and societies represent all of us, in different aspects of our lives, and because life is to that extent inherently paradoxical, there must be some form of compromise amongst them. This is simply unavoidable if we are to protect ourselves not only from states and from markets but, as Gunnar Myrdal observed, from ourselves as well.

If we truly want to address the *global* social question of today—as it appears before us both out there in the world and at home in our own national societies—there is nothing for it, in short, but a judicious balancing of force of state, market, and society. It is the only way we can make any meaningful inroads into our current levels of inequality and avoid the likely future alternatives of an increasingly callous status quo or an eventual bloody confrontation. The peace table already exists, then, and it takes the form of a revivified public realm: the *res publica* of the global age. We are advised to take our seats there before a real crisis comes around.

Both sides of the political spectrum will have their work cut out for them if we are to do this in sufficient numbers. The right will need to get over their dislike of public ownership and socially determined objectives, which are not—as it has been explained to them often enough—the same as nationalisation and collectivisation. And they will need to revisit their faith in the problem-solving potential of the market: something they might begin to do by acknowledging the role the state often plays in their many examples of market triumphs.[24] The right's outdated boosterism and woeful imprecision regarding the foibles of twentieth-century socialism (from which dank well they draw their morals ad infinitum) is little more than intellectual fraud on a grand scale. They must come to the table with their history corrected.

But the left too must overcome its knee-jerk opposition to the private sector: an inconsistent opposition in any case, given that much of the welfare provision that it calls for (such as pensions) relies upon functioning markets. This is not a call for more "third way" solutions, as tried around the world in the 1990s and 2000s (which rather than being open to the private sector tended simply to want to *be* the private sector). It is a call for the left to recognise that corporations are more (and less) than their shareholders and dominant strategies like to claim. The left cannot put all its faith in civil society, or even some "empire" of the movement of movements alone. Creating alternative spaces and challenging hegemonic norms is an important victory, but it is not the final one. The left must come to the table having tempered its Lear-like refusal to demean its noble offices.

With this preamble read out, a more democratic series of negotiations can finally begin. The order of business will remain the same as it was more than seventy years ago at Bretton Woods—a little dog-eared at the edges, but perfectly legible still. The general challenge remains to do what is necessary "so that mankind may be able to meet the urgent challenge of poverty and then move on to better things", as a sharp-eyed observer once put it.[25] The more specific challenge will now be to agree on how we might start to address the underlying, structural forms of injustice that determine both the poverty and suffering of some and the wealth and privileges of others.

And the starting point for all this will be, as Susan Sontag pointed out, to recognise the ways in which our privileges fall upon the same map as others' suffering. Sontag was right about this. But she stood looking *at* the map. We need to enter into it, pen in hand, and set about redrawing it entirely.

NOTES

INTRODUCTION

1 The quintiles compare the average income of these groups and are taken from *World Development Report* (New York: UN Development Programme, 1999), 3. The world Gini indices are from Branko Milanovic, *Global Income Inequality: The Past Two Centuries and Implications for 21st Century* (PowerPoint presentation, Autumn 2011). See also James B. Davies, Anna Sandstrom, Anthony Shorrocks, and Edward N. Wolff, *The World Distribution of Household Wealth* (Discussion Paper No. 2008/03, United Nations University World Institute for Development Economics Research, 2008), esp. 21–22.

2 Credit Suisse Research Institute, *Global Wealth Report 2014* (October 2014), https://www.credit-suisse.com/ch/en/news-and-expertise/research/credit-suisse-research-institute/publications.html. To be counted among the global 1% in terms of wealth requires $798,000 or more in assets at today's prices.

3 This is measured from the baseline year of 1990. The Millennium Development Goals were eight international development goals established at the UN Millennium Summit in 2000 and targeted at the world's poorest, including eradicating extreme poverty and hunger (goal 1) and reducing child mortality (goal 4). The MDGs have been hugely influential in setting the development and aid agenda. In 2015 they were updated with a new set of goals through to 2030, the Sustainable Development Goals (SDGs).

4 Credit Suisse Research Institute, *Global Wealth Report 2014*, 3.

5 According to Branko Milanovic, of the World Bank, these two factors—class and geography—together account for more than 90% of variability in people's global income position. I return to these factors later on. "Where in the World Are You? Assessing the Importance of Circumstance and Effort in a World of Different Mean Country Incomes and (Almost) No Migration" (Policy Research Working Paper No. 4493, World Bank, January 2008). Wealth and quality of life are about more than capital income and wealth, however, on which see also Gerry Kearns and Simon Reid-Henry, "Vital Geographies: Life, Luck, and the Human Condition," *Annals of the Association of American Geographers* 99, no. 3 (2009): 554–74.

6 Credit Suisse Research Institute, *Global Wealth Report 2014*.

7 With due acknowledgment to the late David Foster Wallace, by way of my colleague Ole Jacob Sending.

8 See, for example, David Nally, "Trajectories of Development, Modalities of Enclosure: Land Grabs and the Struggle over Geography," in *At the Anvil:*

Essays in Honour of William J. Smyth, ed. Patrick J. Duffy and William Nolan (Dublin: Geography Publications, 2012), 653–76; see also Ben Bouckley, "Oxfam Tackles PepsiCo, Coke over 'Disastrous Impact' of Sugar Land Grabbing," *FoodNavigator.com*, October 2, 2013.

9 Abhijit Pandya, "Charity Always Begins at Home: House of Lords Report Slam Dunks David Cameron's 'Global Welfare' Project," *Daily Mail*, March 30, 2012. The article refers to "parts of the world that most need to get off their backsides and sort out their nation". Britain is one of the few countries to meet its 0.7% aid contribution.

10 See Jonathan Glennie, "Global Inequality: Tackling the Elite 1% Problem," *Poverty Matters* (blog), November 28, 2011, http://www.theguardian.com/global-development/poverty-matters/2011/nov/28/global-inequality-tackling-elite-national. Glennie gives the figure of £30,000 net annual income in pounds sterling at 2011 prices.

11 Credit Suisse Research Institute, *Global Wealth Report 2014*, 6. On GDP recovery, see Emmanuel Saez, "Striking It Richer: The Evolution of Top Incomes in the United States (Updated with 2009 and 2010 Estimates)," Stanford Center for the Study of Poverty and Inequality, March 2, 2012, http://eml.berkeley.edu/~saez/saez-UStopincomes-2010.pdf, 1.

12 Pierre Rosanvallon, *The Society of Equals* (Cambridge, MA: Harvard University Press, 2013), 10.

13 Richard Titmuss, introduction to *Equality*, by R. H. Tawney (London: Unwin Books, 1964), 10–11.

14 Jean-Jacques Rousseau, *The Social Contract and Discourses*, translated and with introduction by G. D. H. Cole (London: J. M. Dent & Sons, 1966), x.

15 R. H. Tawney, *Equality* (London: Unwin Books, 1964), 42.

CHAPTER 1

1 See Duncan Green, *From Poverty to Power: How Active Citizens and Effective States Can Change the World*, 2nd ed. (Bourton on Dunsmore, UK: Practical Action Publishing, 2012). Green's analysis is a sophisticated one, but global poverty cannot be understood on the basis of a theory of power-as-agency alone.

2 Resilience is widely assumed to be a new concept, for example. Yet this forgets the full title of the Independent Commission's 1980 Brandt Report: "North-South: A Programme for Survival." See also Mark Neocleous, "Commentary: Resisting Resilience," *Radical Philosophy* 178 (March–April 2013): 2–7.

3 Jonathan Glennie, "Andrew Mitchell's Legacy? Aid-Budget Brio That Depoliticized Development," *Poverty Matters* (blog), September 5, 2012, http://www.guardian.co.uk/global-development/poverty-matters/2012/sep/05/andrew-mitchell-legacy-aid-depoliticised-development. Glennie actually

asks, "What kind of politics does the new minister for international development, Justine Greening, espouse?" I hope my paraphrasing here is true to his point, however.

4 "The World's Next Great Leap Forward: Towards the End of Poverty," *Economist*, June 1, 2013. We must also bear in mind the following: "In all, 2.2 billion people lived on less than US$2 a day in 2011, the average poverty line in developing countries and another common measurement of deep deprivation. That is only a slight decline from 2.59 billion in 1981." World Bank, "Poverty Overview," http://www.worldbank.org/en/topic/poverty/overview.

5 Branko Milanovic, "Global Income Inequality by the Numbers: An Overview" (Policy Research Working Paper No. 6259, World Bank, November 2012), http://elibrary.worldbank.org/doi/pdf/10.1596/1813-9450-6259, 12.

6 For all the debate about whether the world is experiencing convergence or divergence, the most likely scenario is one of what we might call "converged-divergence". See also Thomas Piketty, *Capital in the Twenty-First Century* (Cambridge, MA: Belknap Press of Harvard University Press, 2014), 432.

7 See Karl Polanyi, *The Great Transformation: The Political and Economic Origins of Our Time* (1944; Boston: Beacon Press, 2001), 216.

8 Tony Judt, *Ill Fares the Land: A Treatise on Our Present Discontents* (London: Allen Lane, 2010), 138.

9 To take just one example: "The bi-polar distribution of land established during three centuries of colonial rule is still, after nearly two centuries of independence, one of the crucial underpinnings of persistent high levels of income inequality in Latin America." E. H. P. Frankema, "The Colonial Origins of Inequality: The Causes and Consequences of Land Distribution" (Groningen Growth and Development Centre, University of Groningen, June 2006). See also David De Ferranti, Guillermo E. Perry, Francisco H. G. Ferreira, and Michael Walton, *Inequality in Latin America: Breaking with History?* (Washington, DC: World Bank, 2004), https://openknowledge.worldbank.org/handle/10986/15009.

10 My thanks to Alan Lester at the University of Sussex for the always-invigorating discussion and the dispossession examples here.

11 Africa had a median Gini index of 0.42 in the 1990s, for example, compared to Eastern Europe's 0.29. The figure for Central African Republic (from 1993) is 0.61; Ghana's is just 0.37 (figures rounded). See K. Deninger and L. Squire, "New Ways of Looking at Old Issues: Inequality and Growth," *Journal of Development Economics* 57, no. 2 (1998): 259–87; see also Christiania Okojie and Abebe Shimeles, *Inequality in Sub-Saharan Africa: A Synthesis of Recent Research on the Levels, Trends, Effects and Determinants of Inequality in Its Different Dimensions* (London: Overseas Development Institute, February 2006).

12 Ken Henry, "Spreading the Benefits of Globalisation: 'Selling the Compounding Benefits of Reform,'" in *Living Standards and Inequality: Recent Progress and Continuing Challenges*, ed. David Gruen, Terry O'Brien, and Jeremy Lawson (Canberra: Australian Treasury, McMillan Printing Group, 2002), 239–49, 243.

13 Save the Children, *Ending Poverty in a Generation: Save the Children's Vision for a Post-2015 Framework*, (London: Save the Children, 2012), v.

14 John Larkin, "The Wealth Gap," in *Development Asia* (Manila: Asian Development Bank, April 2013), 14.

15 Ibid., 9

16 F. Cingano, "Trends in Income Inequality and Its Impact on Economic Growth" (Social, Employment and Migration Working Paper No. 163, OECD, December 9, 2014), 17, available at http://dx.doi.org/10.1787/5jxrjncwxv6j-en. By modelling the possible effects of historical changes in income inequality on actual 1990–2010 growth rates, the report estimates that "in the United States, the United Kingdom, Sweden, Finland and Norway, the growth rate would have been more than one fifth higher had income disparities not widened. On the other hand, greater equality helped increase GDP per capita in Spain, France and Ireland prior to the crisis" (18).

17 Readers should consult the numerous works by Branko Milanovic outlining the various ways to measure global income inequality. His (mainly Gini-based) approach should also be read alongside that of others. Piketty, for example, holds wealth and income (or labour) measures apart. For Gabriel Palma, the relevant measure of inequality is the ratio of change in income between the top 10% and the bottom 40% (a ratio which mirrors, interestingly enough, Piketty's statistical bracketing of what he terms the "patrimonial middle classes"). The middle 50%, Palma points out, changes relatively little over time compared to the other two groups. See Alex Cobham and Andy Sumner, "Putting the Gini Back in the Bottle? The 'Palma' as a Policy-Relevant Measure of Inequality," https://www.kcl.ac.uk/aboutkings/worldwide/initiatives/global/intdev/people/Sumner/Cobham-Sumner-15March2013.pdf. I raise these issues here as my intention is not to get into the details of income and wealth inequality measures in this book, so much as to draw out and reflect upon the basic trends we might derive from those measures, the interrelationship of those trends (in particular their geographical dynamics), and the macro-political context in which they occur. The figures I draw upon in this paragraph, however, are from http://data.worldbank.org/indicator/SI.POV.GINI; homicide rate comparisons are from https://www.osac.gov/pages/ContentReportDetails.aspx?cid=15771.

18 On the global 1%, see Milanovic, "Global Income Inequality"; and Piketty, *Capital*.

19 These figures are from Odd Arne Westad, *The Global Cold War: Third World Interventions and the Making of Our Times* (Cambridge: Cambridge University Press, 2007), 485.

20 Andrew Sayer, *Why We Can't Afford the Rich* (Bristol, UK: Policy Press, 2015).

21 On net financial assets, see Tax Justice Network, "Revealed: Global Super-Rich Has at Least $21 Trillion Hidden in Secret Tax Havens," press release, July 22, 2012, http://www.taxjustice.net/cms/upload/pdf/The_Price_of_Offshore_Revisited_Presser_120722.pdf, 5. See also the accompanying report by James S. Henry, *The Price of Offshore Revisited* (July 2012), http://www.taxjustice.net/cms/upload/pdf/Price_of_Offshore_Revisited_120722.pdf. On global hunger, see World Food Programme, "10 Things You Need to Know about Hunger in 2013," January 2, 2013, https://www.wfp.org/stories/10-things-you-need-know-about-hunger-2013. On the figure of 2 billion people (2.2 billion for 2011, down, though relative little, from 2.59 billion in 1981), see World Bank, "Poverty Overview," http://www.worldbank.org/en/topic/poverty/overview.

22 Homi Kharas and Geoffrey Gertz, "The New Global Middle Class: A Cross-Over from West to East," in *China's Emerging Middle Class: Beyond Economic Transformation*, ed. Cheng Li (Washington, DC: Brookings Institution Press, 2010), draft version of chap. 2, http://www.brookings.edu/~/media/research/files/papers/2010/3/china%20middle%20class%20kharas/03_china_middle_class_kharas.pdf.

23 My thanks to Danny Dorling for this point.

24 This makes them somewhat different to the rich world's "patrimonial middle classes", as identified by Thomas Piketty (those not in the top 10% but not in the bottom 50% either) in his *Capital*. The existence of the rich world's middle classes may well have been "fragile", as Piketty says (261), but fragility is some way ahead of precarity.

25 See UN Development Programme, *Human Development Report 2013: The Rise of the South: Human Progress in a Diverse World* (New York: UN Development Programme, 2013), 14. See also Martin Ravallion, "The Developing World's Bulging (but Vulnerable) 'Middle Class'" (Policy Research Working Paper No. 4816, World Bank, January 2009). The point about cows is from Stiglitz, who is referring of course to the EU Common Agricultural Policy, which works out at about $2 per cow: Joseph Stiglitz, "The Global Benefits of Equality," *The Guardian*, September 8, 2003, http://www.theguardian.com/environment/2003/sep/08/fairtrade.wto1.

26 The ranking is from Íñigo Moré, *The Borders of Inequality: Where Wealth and Poverty Collide*, trans. Lyn Dominguez (Tucson: University of Arizona Press, 2011), 2.

27 See Gerard Hanlon, *The Dark Side of Management: A Secret History of Management Knowledge* (London: Routledge, 2015).

28 Speaking of the decision to build a new wall between Israel and Egypt to keep out "immigrants and terrorists", Netanyahu told the Israeli daily *Haaretz*, "I took the decision to close Israel's southern border to infiltrators and terrorists. This is a strategic decision to secure Israel's Jewish and democratic character." "Israel to Construct Barrier along Egyptian Border," BBC, January 11, 2010, http://news.bbc.co.uk/2/mobile/middle_east/8451085.stm.

29 The full quote is: "Just as none of us is outside or beyond geography, none of us is completely free from the struggle over geography. That struggle is complex and interesting because it is not only about soldiers and cannons but also about ideas, about forms, about images and imaginings." Edward W. Said, *Culture and Imperialism* (New York: Vintage Books, 1994), 7.

30 See Matthew Sparke and Dimitar Anguelov, "H1N1, Globalization and the Epidemiology of Inequality, *Health and Place* 18 (2012): 726–36.

31 See Gerry Kearns and Simon M. Reid-Henry, "Vital Geographies: Life, Luck and the Human Condition," *Annals of the Association of American Geographers* 99, no. 3 (2009): 554–74.

32 See Mike Davis, *The Monster at Our Door: The Global Threat of Avian Flu* (London: New Press, 2005).

33 Both of these quotes are from "Risk of Deadly TB Exposure Grows along US-Mexico Border," *Wall Street Journal*, March 8, 2013, http://online.wsj.com/news/articles/SB10001424127887323293704578336283658347240.

34 See the recent and, for the IMF, quite remarkable note by Jonathan D. Ostry, Andrew Berg, and Charalambos G. Tsangarides, "Redistribution, Inequality, and Growth" (staff discussion note, International Monetary Fund, April 2014).

35 R. H. Tawney, *Equality*, 5th ed. (London: Unwin, 1964), 90.

36 Cited in Wilkinson and Pickett, *The Spirit Level: Why Equality Is Better for Everyone* (London: Allen Lane, 2009), 55.

37 RMI stands for *revenu minimum d'insertion*, a French minimum subsistence allowance for the unemployed. It is a French variant of workfare.

38 Cited in Stephen Pimpare, *A People's History of Poverty in America* (New York: New Press, 2011), 3.

39 This is an argument made in Gareth Stedman-Jones, *An End to Poverty: A Historical Debate* (London: Profile Books, 2004).

40 "Becoming deference" was what the Committee for Improving the Condition of Free Blacks required from those seeking their aid in the late eighteenth century. Pimpare, *People's History*, 4.

41 Ibid., 7.

42 George R. Boyer, review of Lynn Hollen Lees, "The Solidarities of Strangers: The English Poor Laws and the People, 1700–1948" (Cambridge: Cambridge University Press, 1998), *H-Net Reviews*, http://www.h-net.org/reviews/showrev.php?id=4162.

43 Michael Ignatieff, *The Needs of Strangers* (New York: Picador, 1984), 17.

44 Gunnar Myrdal, *The Challenge of World Poverty* (New York: Pantheon Books, 1970), 8.

45 Wilkinson and Pickett, *Spirit Level*, 60; see also the World Bank SIMA database (1998), where the standout countries are Denmark, Norway, Sweden, the Netherlands, and Portugal; see also Alessandra Cepparulo and Luisa Giuriato, "Aid Financing of Global Public Goods: An Update" (Munich Personal RePEc Archive Paper No. 22625, May 11, 2010), 28.

46 UN Economic Commission for Africa, *African Economic Report* (1999).

47 Kwame Anthony Appiah, *Cosmopolitanism: Ethics in a World of Strangers* (London: Penguin, 2007), 172.

48 Among the few exceptions are the UN Development Programme's *Human Development Reports* from 1999 and 2005. See also Yuri Dikhanov, *Trends in Global Income Distribution, 1970–2000 and Scenarios for 2015* (occasional paper, Human Development Report Office, 2005).

CHAPTER 2

1 Susan Sontag, *Regarding the Pain of Others* (New York: Farrar, Straus & Giroux, 2003), 102–3.

2 Given that Leibovitz was longtime partner of Susan Sontag, the image has a certain irony appended to it in our context.

3 See Kaitlyn Bonneville, "Louis Vuitton Core Values Campaign Supports Multiple Charities," *Luxury Daily*, September 28, 2010, http://www.luxurydaily.com/louis-vuitton-core-values-campaign-supports-multiple-charities/.

4 Ibid.

5 For various sources on this, see Richard Tomlinson and Fergal O'Brien, "Bono Likes to Preach but Hates Tax," *Sydney Morning Herald*, January 29, 2007, http://www.smh.com.au/news/business/bono-likes-to-preach-but-hates-tax/2007/01/28/1169919210561.html. For the Tax Justice Network's work, see their website at http://www.taxjustice.net.

6 These figures are from the Eurodad report *G8 Debt Deal One Year On: What Happened and What's Next?* (June 2006), http://eurodad.org/318/?lang=es. The report contains a useful overview of both the initial claims for $40 billion of relief over forty years (after prerequisites were fulfilled) and its achievements one year on.

7 Cited in Duncan Green, *From Poverty to Power: How Active Citizens and Effective States Can Change the World*, 2nd ed. (Bourton on Dunsmore, UK: Practical Action Publishing, 2012), 14.

8 William Beveridge, *Social and Allied Services* (London: His Majesty's Stationery Office, 1942), 6. The quote is a well-remembered one. What is less often recalled is that it was immediately followed by a second "guiding principle":

social insurance should be just one part of a "comprehensive policy of social progress". The point on comprehensiveness is salient. Beveridge identified "five giants" on the road to reconstruction (all of which are more familiar today from the debate as it pertains to poor countries: want, disease, ignorance, squalor, and idleness. The Beveridge Report can be viewed in full at http://news.bbc.co.uk/2/shared/bsp/hi/pdfs/19_07_05_beveridge.pdf.

9 Lewis Coser, "The Sociology of Poverty: To the Memory of Georg Simmel," *Social Problems* 14, no. 2 (1965): 141.

10 Akhil Gupta, "The Construction of the Global Poor: An Anthropological Critique," *World Social Science Report 2010: Knowledge Divides* (Paris: UNESCO, 2010), secs. 1.1, 13–16. For the changing idea of poverty during the nineteenth century, see in particular Gertrude Himmelfarb, *The Idea of Poverty: England in the Early Industrial Age* (London: Faber and Faber, 1985); and Gareth Stedman Jones, *An End to Poverty? A Historical Debate* (London: Profile Books, 2004).

11 Thomas Pogge, "Growth and Inequality: Understanding Recent Trends and Policy Choices," *Dissent* 55, no. 1 (2008): 66–75.

12 "Failed consumers" is Zygmunt Bauman's phrase in *Wasted Lives: Modernity and Its Outcasts* (Cambridge, UK: Polity Press, 2004), 5.

13 Akhil Gupta explains the difference as clearly as any: "PRSPs are country-driven, result-oriented strategies that bring national development plans in line with neoliberal globalization by emphasizing growth, free markets and an open economy. . . . However, they differ from structural adjustment programmes through their emphasis on the need for broad-based growth strategies, good governance, decentralization, empowerment, investments in health care, education and human capital, and social protection for those who are adversely affected by adjustment processes." "The Construction of the Global Poor: An Anthropological Critique," in *World Social Science Report 2010: Knowledge Divides* (Paris: UNESCO, 2010), 14.

14 On the matter of the billionaires, see Thomas Piketty, *Capital in the Twenty-First Century* (Cambridge, MA: Belknap Press of Harvard University Press, 2014), 433.

15 Oxfam, "The Cost of Inequality: How Wealth and Income Extremes Hurt Us All" (media briefing, January 18, 2013).

16 Jeffrey Sachs, *Common Wealth: Economics for a Crowded Planet* (London, Penguin, 2009), 6.

17 Fortunately, thanks to the IF campaign, we can do more than just think of this. Search for the video "What Has Aid Ever Done for Anyone?" at YouTube.

18 See William Easterly, *The White Man's Burden: Why the West's Efforts to Aid the Rest Have Done So Much Ill and So Little Good* (Oxford: Oxford University Press, 2007), 41–44. Easterly is even more specific about the negative impact

of state powers and would-be saviours in his latest book, *The Tyranny of Experts: Economists Dictators and the Forgotten Rights of the Poor* (New York: Basic Books, 2013). But Easterly, I would contend, is wrong. What is really at work in this liberal approach to the administration of poverty is, first and foremost, a simultaneous withdrawal of state support from a wide variety of the state's former areas of responsibility, along with a simultaneous reconfiguration of the idea of society upon which that state responsibility is extended, one that holds individuals themselves responsible for making up the difference. At the same time, the power of the state has grown to enforce the new logic and above all to pin down and hold in place the economic system it corresponds to. To wit, states have become both more powerful and less responsible. It is this that explains the phenomena Easterly takes to be indicative of the problem of 'the state', not the ideological orientation of those statesmen he singles out for criticism.

19 Abhijit V. Banerjee and Esther Duflo, *Poor Economics: Barefoot Hedge-Fund Managers, DIY Doctors and the Surprising Truth about Life on Less Than $1 a Day* (London: Penguin, 2012), 206.

20 Ha-Joon Chang, comments made at a conference at Cambridge University. The tenor of response to Chang's arguments is worth reading. See Ha-Joon Chang, "Reply to the Comments on 'Institutions and Economic Development: Theory, Policy, and History,'" *Journal of Institutional Economics* 7, no. 4 (2011): 595–613.

21 Rafael La Porta, Florencio Lopez-de-Silanes, Andrei Shleifer, and Robert Vishny, "The Quality of Government," (Working Paper No. 6727, National Bureau of Economic Research, Cambridge, MA, September 1998), 2, http://www.nber.org/papers/w6727.

22 Daron Acemoglu and James A. Robinson, *Why Nations Fail: The Origins of Power, Prosperity and Poverty* (London: Profile Books, 2012), 3.

23 Ha-Joon Chang, "Institutions and Economic Development: Theory, Policy and History," *Journal of Institutional Economics* 7, no. 4 (2011): 473–98; see also his "Reply."

24 The Marx quote is from Catherine Hall's scintillating lecture "Gendering Property, Racing Accumulation" (presented at History after Hobsbawm conference, London's Senate House, April 29–May 1, 2014). The original is from *Capital*, vol. 1, chap. 31. See also Catherine Hall, Nicholas Draper, Keith McClelland, Katie Donington, and Rachel Lang, *Legacies of British Slave Ownership: Colonial Slavery and the Formation of Victorian Britain* (Cambridge: Cambridge University Press, 2014), 9–12; and, of course, Eric Williams, *Capitalism and Slavery* (1944; Chapel Hill: University of North Carolina Press, 1994), which first outlined the role of slavery as a key *contributor* to British industrial development.

25 On the poverty data, see Atul Kohli, *Poverty amid Plenty in the New India* (Cambridge: Cambridge University Press, 2012), 3. The GDP growth figures are from the World Bank: "GDP Growth (Annual %)," http://data.worldbank.org/indicator/NY.GDP.MKTP.KD.ZG.

26 The phrase is of course Colin Clark's, in his *Conditions of Economic Progress* (1940). Clark is sometimes seen as the grandfather of the idea of economic growth. In Raymond Aron's view, Clark was the éminence grise behind the doctrine of economic growth. And yet it was Keynes who secured Clark one of his first teaching posts and cited him in *The Economic Consequences of the Peace*, whereas Clark's own formulation of growth (that, as stated here, it produces capital, not the other way round) led him to see the need to promote education as central. Tawney would not have been entirely in disagreement. See Raymond Aron, *The Dawn of Universal History: Selected Essays from a Witness of the Twentieth Century*, trans. Barbara Brey, ed. Yair Reiner (New York: Basic Books, 2008).

27 Anyone interested in this should read the articles by Robert Hunter Wade, of the London School of Economics. Among those which have been most influential in my own thinking are "Is Globalization Reducing Poverty and Inequality?" *World Development* 32, no. 4 (2004): 567–89, and "Globalization, Growth, Poverty and Inequality," in *Global Political Economy*, ed. John Ravenhill, 4th ed. (Oxford: Oxford University Press, 2014), 305–43.

28 See, for example, among various sources of evidence against this claim, Santosh Mehrotra, *Integrating Economic and Social Policy: Good Practices from High-Achieving Countries* (Innocenti Working Paper No. 80, UNICEF Innocenti Research Centre, Florence, Italy, 2000).

29 Tim Jackson, *Prosperity without Growth: Economics for a Finite Planet* (London: Earthscan, 2009).

30 See David Dollar and Aart Kraay, "Growth Is Good for the Poor" (Policy Research Working Paper No. WPS 2587, World Bank, 2000), http://siteresources.worldbank.org/DEC/Resources/22015_Growth_is_Good_for_Poor.pdf, 11.

31 Of the various places Harvey discusses this, see his *The Enigma of Capital* (London: Profile Books, 2011).

32 Michael Rowan, "We Need to Talk about Growth (and We Need to Do the Sums as Well)," January 21, 2014, *Persuade Me*, http://persuademe.com.au/need-talk-growth-need-sums-well/. See also H. E. Daly, *Steady-State Economics* (Washington, DC: Island Press, 1991).

33 Hyun H. Son and Nanak Kakwani, "Global Estimates of Pro-Poor Growth," *World Development* 36, no. 6 (2008): 1048–66.

34 Nancy Birdsall, "The World Is Not Flat: Inequality and Injustice in Our Global Economy" (WIDER Annual Lecture No. 9, March 2006, UNU-WIDER), http://www.wider.unu.edu/publications/annual-lectures/en_GB/AL9/.

35 Paul Collier, *The Bottom Billion: Why the Poorest Countries Are Failing and What Can Be Done about It* (Oxford: Oxford University Press, 2007), 157.

36 For an insightful analysis of the problems with Collier's methodology, see Peter Lawrence, "Development by Numbers," *New Left Review*, March–April 2010, 143–53.

37 The Sen observation is from Marc Wuyts, "Inequality and Poverty as the Condition of Labour" (draft paper for "The Need to Rethink Development Economics," UNRISD conference, Cape Town, South Africa, 2001), 5–6.

38 David Leonhardt, "NYU Lands Top Economist for Cities Project," *New York Times*, May 27, 2011.

39 Keane Bhatt, "Reporting on Romer's Charter Cities: How the Media Sanitize Honduras' Brutal Regime," NACLA, https://nacla.org/news/2013/2/19/reporting-romer's-charter-cities-how-media-sanitize-honduras's-brutal-regime.

40 See also David Kirkpatrick, "Vinod Khosla's Ingenious War on Global Poverty," *Daily Beast*, October 21, 2010, http://www.thedailybeast.com/articles/2010/10/22/vinod-khoslas-ingenious-war-on-global-poverty.html.

41 Jason Miklian and Kristian Hoelscher, "A Tale of New Cities," *Harvard International Review* 35, no. 4 (Spring 2014): 13–18, http://hir.harvard.edu/archives/5811.

42 In a recent World Bank paper Romer and coauthor Brandon Fuller call for a new era of market-run cities on the following terms, for example. "New cities can compete for residents," they say, "allowing countries to use new political entities to try different types of rules, subjecting them to a market test of opt-in that can operate alongside the more familiar test of voice." "Urbanization as Opportunity" (Policy Research Paper No. 6874, World Bank, May 2014), 9.

CHAPTER 3

1 This is the title of Michael Latham's excellent book about modernisation: *The Right Kind of Revolution: Modernization, Development, and U.S. Foreign Policy from the Cold War to the Present* (Ithaca, NY: Cornell University Press, 2011).

2 Enrique Dussel, "Eurocentrism and Modernity: An Introduction to the Frankfurt Lectures," *Boundary 2* 20, no. 3 (1993): 65–76.

3 Walt W. Rostow, *The Stages of Economic Growth: A Non-Communist Manifesto* (Cambridge: Cambridge University Press, 1960).

4 Cited in Walt W. Rostow, *Concept and Controversy: Sixty Years of Taking Ideas to Market* (Austin: University of Texas Press, 2010), 215.

5 Eric Helleiner, "Reinterpreting Bretton Woods: International Development and the Neglected Origins of Bretton Woods," *Development and Change* 37, no. 5 (2006): 943–67, 963–64.

6 Cited in David Engerman, "The Romance of Economic Development and New Histories of the Cold War," *Diplomatic History* 28, no. 1 (2004): 30.
7 Cited in Mark Mazower, *Governing the World: The History of an Idea* (London: Allen Lane, 2012): 263.
8 Garrett Hardin, "The Tragedy of the Commons," *Science* 162 (3859): 1243–48.
9 See Erez Manela, "Roundtable on Nick Cullather's 'The Hungry World,'" *Passport*, January 2012, 6.
10 Cited in Michele Alacevich, *The Political Economy of the World Bank: The Early Years* (Stanford, CA: Stanford University Press, 2009), 71.
11 See Adrian Guelke, "African Socialism in Two Countries by Ahmed Mohiddin; Uganda: A Modern History by Jan Jelmert Jorgensen," *Third World Quarterly* 4, no. 3 (1982): 559–61. It is interesting also to compare Kenya's *African Socialism and Its Application to Planning in Kenya* (1965) to Tanzania's Arusha Declaration (1967) to Obote's Common Man's Charter. Kenya was more conservative and Tanzania more radical, yet all three took the same colonial inheritance as their starting point.
12 Odd Arne Westad, "The Project," *London Review of Books* 30, no. 2 (January 24, 2008): 30–31.
13 Thandika Mkandawire, "Thinking about Developmental States in Africa," *Cambridge Journal of Economics* 25, no. 3 (May 2001): 289–313.
14 Cited in Thandika Mkandawire, "From the National to the Social Question," *Transformation: Critical Perspectives on Southern Africa*, no. 69 (2009): 134.
15 Cited in Engerman, "Romance of Economic Development," 36.
16 Fidel Castro, "To Create Wealth with Social Conscience," in *Man and Socialism in Cuba: The Great Debate*, ed. Bertram Silverman (New York: Atheneum, 1971).
17 Karl Polanyi, *The Great Transformation: The Political and Economic Origins of Our Time* (1944; Boston: Beacon Press, 2001), xxvii.
18 Jimi O. Adesina, "In Search of Inclusive Development: Social Policy in Sub-Sahara Africa Context," in *Social Policy in Sub-Saharan African Context: In Search of Inclusive Development* (Basingstoke, UK: Palgrave, 2007), 11.
19 Gunnar Myrdal, *The Challenge of World Poverty: A World Anti-Poverty Programme in Outline* (New York: Pantheon Books, 1970), 60.
20 On per capita income of developed countries and mortality rates, see Hla Mlint, "Comment," in *Pioneers in Development*, ed. Gerald M. Meier and Dudley Seers, A World Bank Publication (Oxford: Oxford University Press, 1985), 1:167. On the continent's own "golden age", see Adesina, "In Search of Inclusive Development," 3. On the eight largest countries, see Alicia Puyana Mutis, "Pondering the Hurdles for the Mexican Economy While Reading *Capital in the Twenty-First Century*," *Real-World Economics Review* 69 (2014): 74–89, 78. On the thirty fastest-growing economies, see World Bank Development Indicators, 1998, cited in Mkandawire, "Thinking," 304.

21 Eric Helleiner, *States and the Reemergence of Global Finance: From Bretton Woods to the 1990s* (Ithaca, NY: Cornell University Press), 25.
22 John Gerard Ruggie, "International Regimes, Transactions, and Change: Embedded Liberalism in the Postwar Economic Order," *International Regimes* 36, no. 2 (1982): 379–415. The quote itself is from Helleiner, *States*, 3.
23 Helleiner, "The Southern Side of Embedded Liberalism: The Politics of Postwar Monetary Policy in the Third World (Working Paper No. 01/5, Trent International Political Economy Centre), 1.
24 Ibid., 5.
25 Cited in Giuliano Garavini, "The Colonies Strike Back: The Impact of the Third World on Western Europe, 1968–1975," *Contemporary European History* 16, no. 3 (2007): 301.
26 Ibid., 316.
27 Ibid, 317.
28 Helleiner, "Southern Side," 6.
29 Sheyda Jahanbani, "One Global War on Poverty: The Johnson Administration Fights Poverty at Home and Abroad, 1964–1968," in Beyond the Cold War: Lyndon Johnson and the New Global Challenges of the 1960s, ed. Francis J. Gavin and Mark A. Lawrence (New York: Oxford University Press, 2014), 97–117.
30 Sharad Chari and Katherine Verdery, "Thinking between the Posts: Postcolonialism, Postsocialism, and Ethnography after the Cold War," *Comparative Studies in Society and History* 51, no. 1 (2009): 6–34.
31 Amy Saywar, "An Intellectual History of Development," *Passport*, January 2012, 9–11.
32 Myrdal, *Challenge of World Poverty*, 159.
33 See Stephen Gill, *American Hegemony and the Trilateral Commission* (Cambridge: Cambridge University Press, 1991).
34 See Mark Mazower, *Governing the World: The History of an Idea* (London: Allen Lane, 2012), who makes this argument with particular clarity (see especially chap. 12).
35 Vijay Prashad, *The Poorer Nations: A Possible History of the Global South* (London: Verso, 2013), 45.
36 Myrdal, *Challenge of World Poverty*, 309.
37 Ibid.
38 Jubilee Debt Campaign, *The State of Debt: Putting an End to Thirty Years of Crisis* (May 20, 2012), 2.
39 David Harvey, *Spaces of Global Capitalism: Towards a Theory of Uneven Geographical Development* (London: Verso, 2006).
40 Prashad, *Poorer Nations*, 52.
41 The London Inter Bank Offered Rate (LIBOR) rate nearly doubled from 1978 to 1981. See Stuart Corbridge, *Debt and Development* (London: Blackwell, 2002), 38.

42 Cited in Tony Judt, *Ill Fares the Land: A Treatise on Our Present Discontents* (London: Allen Lane, 2010), 78.
43 Corbridge, *Debt and Development*, 49.
44 See Odd Arne Westad, *The Global Cold War: Third World Interventions and the Making of Our Times* (Cambridge: Cambridge University Press, 2007), 360.
45 Corbridge, *Debt and Development*, 60.
46 Adesina, "In Search of Social Development," 2.
47 Ibid., 12.
48 Prashad, *Poorer Nations*, 81.
49 James M. Scott, ed., *Deciding to Intervene: The Reagan Doctrine and American Foreign Policy* (Durham, NC: Duke University Press, 1996), 2.
50 Ibid., 4. Ultimately there was no uprising in Mozambique.
51 See Angus Bergin, *The Great Persuasion: Reinventing Free Markets since the Depression* (Cambridge, MA: Harvard University Press, 2012).
52 For a longer discussion of global conservatism, see Will Hutton, *The World We're In* (London: Abacus, 2003), 7–24.
53 See Prashad, *The Darker Nations: A People's History of the Third World* (New York: New Press, 2007), 64–70.
54 Cited in Corbridge, *Debt and Development*, 100.
55 This phrase is the IMF's own: "IMF, African Prospects Tied to Courageous Adjustment Efforts," *IMF Survey* 25, no. 15 (1996): 259.
56 Cited in Derek Malcolm, "Tomas Gutierrez Alea: Memories of Underdevelopment," *The Guardian*, February 10, 2000, http://www.guardian.co.uk/film/2000/feb/10/artsfeatures.
57 Craig Calhoun, "The Idea of Emergency: Humanitarian Action and Global (Dis)order," in *Contemporary States of Emergency: The Politics of Military and Humanitarian Interventions*, ed. Didier Fassin and Mariella Pandolfi (Cambridge, MA: Zone Books, 2011), 49.
58 For an intriguing argument showing how radicals are no less prone to regurgitating some of the steeper forms of ignorance about the world, see Matthew Sparke, "Everywhere but Always Somewhere: Critical Geographies of the Global South," *Global South* 1, no. 1 (2007): 117–26. The actual Ignatieff quote is "Doctrines are words, and whips are things," from *The Needs of Strangers* (1984; New York: Picador, 2001), 52.
59 See Nicolas Guilhot, *The Democracy Makers: Human Rights and the Politics of Global Order* (New York: Columbia University Press, 2005), 223.
60 Samuel Moyn, "Totalitarianism, Famine and Us," *The Nation*, November 7, 2012, http://www.thenation.com/article/171122/totalitarianism-famine-and-us.
61 Padma Desai, "The Soviet Union and Cancun," *Third World Quarterly* 4, no. 3 (1982): 514.

62 Samuel Moyn, *The Last Utopia: Human Rights in History* (Cambridge, MA: Harvard University Press, 2012). See also Moyn, *The Powerless Companion: Human Rights in the Age of Global Inequality* (Cambridge, MA: Harvard University Press, forthcoming).

63 See John Rawls, "Four Lectures on Henry Sidgwick," in *John Rawls Lectures on the History of Political Philosophy*, ed. Samuel Freeman (Cambridge MA: Belknap Press of Harvard University Press, 2007), 393.

64 "One-third of all developing country governments—more than 40 governments in all—entered the 1990s owing on average 220% of their gross domestic product (GDP) to foreign lenders." Thomas Oatley, "Political Institutions and Foreign Debt in the Developing World," *International Studies Quarterly* 54 (2010): 175–95.

65 Giovanni Andrea Cornia, Richard Jolly, and Frances Stewart, eds., *Adjustment with a Human Face: Protecting the Vulnerable and Promoting Growth* (New York: Oxford University Press, 1987).

66 This is a point made by Richard Peet, *Geography of Power: The Making of the Global Economic System* (London: Zed Books, 2007), 48.

67 See Jean-Philippe Therien, "Beyond the North-South Divide: The Two Tales of World Poverty," *Third World Quarterly* 20, no. 4 (August 1999): 723–42, 730.

68 See Westad, *Global Cold War*, 397–400.

69 James D. Wolfensohn, "Choosing a Better World", *Asia Times*, January 22, 2003, http://atimes.com/atimes/Global_Economy/EA22Dj01.html.

70 Moyn, *Powerless Companion*; see also Katrina Forrester, "Citizenship, War, and the Origins of International Ethics in American Political Philosophy, 1960–1975," *Historical Journal* 57, no. 3 (2014): 773–801.

CHAPTER 4

1 See Delly Mawazo Sesete, "Apple: Time to Make a Conflict Free iPhone," *Comment Is Free* (blog), http://www.theguardian.com/commentisfree/cifamerica/2011/dec/30/apple-time-make-conflict-free-iphone. Apple more recently (in February 2015) said that it had audited all the smelters it uses in its supply chain and none of them uses minerals from conflict regions. But Apple has not divulged its auditing procedures, and there are loopholes that could still allow conflict minerals into its supply chain. Lynnley Browning, "Where Apple Gets the Tantalum for Your iPhone," *Newsweek*, February 4, 2015, *http://www.newsweek.com/2015/02/13/where-apple-gets--tantalum-your-iphone-304351.html*. Moreover, the wider moral question does not hinge on the supply chain issue alone.

2 See Gerry Kearns, "Progressive Geopolitics," *Geography Compass* 2, no. 5 (2008): 1599–1620.

3 Nicholas Watt, "David Cameron Calls for UK Arms Sales to India," *The Guardian*, April 11, 2012, http://www.guardian.co.uk/politics/2012/apr/11/david-cameron-trade-mission-indonesia.

4 On the arms trade, see Rachel Stohl, "Coordinating a Global Strategy for the International Arms Trade," International Relations and Security Network (ISN) Digital Library, August 21, 2014; see also the numerous publications of the Small Arms Survey network under Keith Krause in Geneva. See in particular "Reducing Illicit Arms Flows and the New Development Agenda" (Research Notes No. 50, Small Arms Survey, March 2015), 1. For the Eisenhower quote, see Richard Jolly, "Disarmament for Human Development: Vision and Challenge for the Twenty-First Century" (panel discussion at the United Nations, October 21, 1997), http://disarm.igc.org/2009backup/T211097humandev.html.

5 D. Rodrik and A. Subramanian, "Why Did Financial Globalization Disappoint?" (Staff Paper No. 56, International Monetary Fund, January 2009).

6 Ray Kiely, *The Clash of Globalisations: Neoliberalism, the Third Way, and Anti-Globalisation* (Boston: Brill, 2005), 103–6.

7 Cited in ibid., 105

8 See Joe Murphy, "Blair Pushed through Deal for Indian Billionaire Who Gave Labour £125,000," *The Telegraph*, February 10, 2002.

9 Felicity Lawrence, "Quarter of FTSE 100 Subsidiaries Located in Tax Havens," *The Guardian*, October 11, 2011, http://www.theguardian.com/business/2011/oct/11/ftse-100-subsidiaries-tax-havens.

10 ActionAid, *Collateral Damage* (March 2012), http://www.actionaid.org.uk/sites/default/files/doc_lib/collateral_damage.pdf, 2.

11 Peter Eigen, "A Prosperous Africa Benefits Everybody," *Journal of World Energy Law and Business* 7, no. 1 (2014): 6.

12 See L. Ndikumana and J. K. Boyce, *Africa's Odious Debts: How Foreign Loans and Capital Flight Bled a Continent* (London: Zed Books), cited in Jubilee Debt Campaign, *The State of Debt: Putting an End to Thirty Years of Crisis* (May 20, 2012), 11.

13 Dani Rodrik, "The Nation State Reborn," *Project Syndicate*, February 13, 2012, http://www.project-syndicate.org/commentary/the-nation-state-reborn.

14 J. Agnew, "The Territorial Trap: The Geographical Assumptions of International Relations Theory," *Review of International Political Economy* 1, no. 1 (1994): 53–80.

15 See Kristin Bergtora Sandvik, "Fighting the War with the Ebola Drone," *Norwegian Centre for Humanitarian Studies* (blog), December 3, 2012, http://www.humanitarianstudies.no/2014/12/03/fighting-the-war-with-the-ebola-drone/; and Paul Farmer, "Diary," *London Review of Books* 36, no. 20 (2014): 38–39.

16 Thomas P. M. Barnett, *The Pentagon's New Map: War and Peace in the Twenty-First Century* (New York: G. P. Putnam's Sons, 2004).

17 See Branko Milanovic, "Global Income Inequality by the Numbers: In History and Now" (Policy Research Working Paper 6259, World Bank, November 2012). Elsewhere he has it as 90%. See "Where in the World Are You? Assessing the Importance of Circumstance and Effort in a World of Different Mean Country Incomes and (Almost) No Migration" (Policy Research Working Paper No. 4493, World Bank, January 2008).

18 See S. M. Reid-Henry, "An Incorporating Geopolitics: Frontex and the Geopolitical Rationalities of the European Border," *Geopolitics* 18, no. 1 (2012): 198–224.

19 Alan Travis, "Detention Centre Castigated over Death of Elderly Man," *The Guardian*, January 16, 2014, http://www.theguardian.com/uk-news/2014/jan/16/harmondsworth-elderly-man-died-handcuffs.

20 See Jonathan Weisman, "Trans-Pacific Partnership Seen as Door for Foreign Suits against U.S.," *New York Times*, March 25, 2015, http://www.nytimes.com/2015/03/26/business/trans-pacific-partnership-seen-as-door-for-foreign-suits-against-us.html. For NAFTA on steroids, the group in question is Global Trade Watch, which is cited prominently at the Stop TPP campaign website, at stoptpp.org (the campaign is a case of Occupy actually turning its attention to the international level).

21 See, for example, Mary Kaldor, "Civil Society and Accountability" *Journal of Human Development* 4, no. 1 (2003): 5–27. The phrase is mine.

22 Nicolas Guilhot, *The Democracy Makers: Human Rights and the Politics of Global Order* (New York: Columbia University Press, 2005), 190.

23 Cited in Michele Alacevich, *The Political Economy of the World Bank: The Early Years* (Stanford, CA: Stanford University Press, 2009), 129. The comment being Vice-President Robert Garner's to Lauchlin Currie, in 1951.

24 A transcript of Tom Ridge's swearing-in ceremony is available at "America Strikes Back: Tom Ridge Worn in as New Director of Homeland Security," CNN, October 8, 2001, http://transcripts.cnn.com/TRANSCRIPTS/0110/08/se.09.html.

25 Mark Neocleous, "From Social to National Security: On the Fabrication of Economic Order," *Security Dialogue* 37 (2006): 363–84, 367.

26 Ibid., 376.

27 Ibid.

28 Cited in Esping-Andersen, *The Three Worlds of Welfare Capitalism* (Cambridge: Polity Press, 1989), 33.

29 Brad Evans, "Terror in All Eventuality," *Theory & Event* 13, no. 3 (2010): 1.

30 Jean Michel Severino and Olivier Ray, "The End of ODA: Death and Rebirth of a Global Public Policy" (Working Paper No. 167, Center for Global Development, Washington, DC, March 2009), 3n4.

31 Patrick Wintour, "Peers Warn of Terror Bill Cuts," *The Guardian,* November 28, 2001, http://www.theguardian.com/politics/2001/nov/28/uk.september11.

32 Rein Mullerson, "Book Review of Koskenniemi: *The Gentle Civiliser of Nations,*" *European Journal of International Law* 13 (2002): 725–35.

33 Martti Koskenniemi, "International Law and Hegemony: A Reconfiguration," *Cambridge Review of International Affairs* 17, no. 2 (2004): 197–218.

34 See Håvard Hegre, "Peace on Earth: The Future of Internal Armed Conflict," *Significance, Journal of the Royal Statistical Society* 4 (2013): 4–8.

35 The full text of the ILO's constitution is available at http://www.ilo.org/dyn/normlex/en/f?p=1000:62:0::NO:62:P62_LIST_ENTRIE_ID:2453907:NO.

36 Jonathan Glennie, "Is It Time for Mali to Plan an Exit Strategy from Aid?" (speech to the Annual Retreat of Technical Partners and Financiers, Bamako, February 8, 2011), http://www.odi.org/sites/odi.org.uk/files/odi-assets/publications-opinion-files/7149.pdf. He is here citing the work of Paolo de Renzio and Joseph Hanlon, "Contested Sovereignty in Mozambique: The Dilemmas of Aid Dependence" (Global Economic Governance Programme Working Paper No. 2007/25, Managing Aid Dependency Project, University College, Oxford, 2007).

37 Alexander Downes, cited in Patricia Owens, "Accidents Don't Just Happen: The Liberal Politics of High-Technology 'Humanitarian' War," *Millennium: Journal of International Studies* 32, no. 3 (2003): 595–616.

38 Patrick Hayden, "Superfluous Humanity: An Arendtian Perspective on the Political Evil of Global Poverty," *Millennium—Journal of International Studies* 35 (2007): 279–300.

39 The phrase is Foucault's (channelling La Perrière): See Graham Burchell, Colin Gordon, and Peter Miller, eds., *The Foucault Effect: Studies in Governmentality* (Chicago: University of Chicago Press, 1991), 93.

40 Amartya Sen, "Equality of What?" (Tanner Lectures on Human Values, Stanford University, May 22, 1979), http://tannerlectures.utah.edu/_documents/a-to-z/s/sen80.pdf.

CHAPTER 5

1 Karl Polanyi, *The Great Transformation: The Political and Economic Origins of Our Time* (1944; Boston: Beacon Press, 2001).

2 Göran Therborn, *The Killing Fields of Inequality* (Cambridge: Polity Press, 2013), 183.

3 See *World Investment Report 2014: Investing in the SDGs. An Action Plan* (Geneva: UNCTAD, 2014). On the MDG shortfall, see William Minter, "Global Solidarity Levy Urgently Needed," *Epoch Times,* September 19, 2010, http://www.theepochtimes.com/n2/opinion/global-solidarity-levy-urgently-needed-42866.html.

4 Albert O. Hirschmann, *A Bias for Hope: Essays on Development and Latin America* (New Haven, CT: Yale University Press, 1971). His lesson there is not only that "economic and political forces interact" (1) but that they go on doing so, even after ideal-type arrangements have been advocated or put into place. The implication is that this needs building in a priori and that allowance for (democratic) iterations of any original structure is part of the challenge of conceiving of that very structure in the first place. We are all duly warned. But as Hirschmann goes on to counsel, in a manner reflected in this book, the next best thing to having something is participating in the process of bringing it about (which begins, of course, with imagining it) (6).

5 Kristian Stokke and Olle Törnquist, eds., *Democratization in the Global South: The Importance of Transformative Politics* (Basingstoke, UK: Palgrave Macmillan, 2013). As Peter Baldwin points out, this was one history among many others, and there are no one-size-fits-all models to be derived from it. In every case, the social bases of solidarity vary. But there are dynamics to be gleaned. See Peter Baldwin, *The Politics of Social Solidarity: Class Bases of the European Welfare State 1875–1975* (Cambridge: Cambridge University Press, 1993), 289–92. The term "project" here is a deliberate echo of Vijay Prashad's use of the term to describe the radical but reformist wing of Third Worldism, as institutionalised in the Non-Aligned Movement. *The Darker Nations: A People's History of the Third World* (London: Verso, 2008).

6 Sheri Berman, "Social Democracy and the Creation of the Public Interest," *Critical Review: A Journal of Politics and Society* 23, no. 3 (2011): 237–56. On governments that came and went, see Francis Sejerstad, *The Age of Social Democracy: Norway and Sweden in the Twentieth Century*, trans. Richard Daly (Princeton, NJ: Princeton University Press, 2011), 73.

7 See Donald Sassoon, *One Hundred Years of Socialism: The West European Left in the Twentieth Century*, rev. ed. (London: I. B. Tauris, 2010). Sassoon also rightly notes the root of social democracy's present malaise in its turn, at point of inception, away from internationalism (42–47).

8 It is true that the economic upswing referred to here had international roots, but the point was that the Scandinavian countries were able to make the most of that upswing while others were not.

9 Stokke and Törnquist, *Democratization*, 31–32.

10 Sejerstad, *Age of Social Democracy*, 53.

11 Ibid., 106–14.

12 Frank Fischer and Gerald J. Miller, *Handbook of Public Policy Analysis: Theory, Politics, Methods* (Boca Raton, FL: CRC Press, 2006), 341. See also Patricia P. Martin and David A. Weaver, "Social Security: A Program and Policy History," *Social Security Bulletin* 66, no. 1 (2005), http://ssrn.com/abstract=2121776.

13 Sejerstad, *Age of Social Democracy*, 100.

14 On Sweden, see ibid., 35.

15 See Olli Kangas and Joakim Palme, *Social Policy and Economic Development in the Nordic Countries* (Basingstoke, UK: Palgrave Macmillan and UNRISD, 2005), 2–3; see also Olli Kangas and Joakim Palme, "Making Social Policy Work for Economic Development: The Nordic Experience," *International Journal of Social Welfare* 18 (2009): S62–S72.

16 Cited in Sejerstad, *Age of Social Democracy*, 121.

17 Norwegian Ministry of Finance, "Women in Work: The Norwegian Experience," *OECD Observer*, no. 239 (November 2012).

18 Sejerstad, *Age of Social Democracy*, 122.

19 Ibid., 111.

20 Ann Pettifor, *The Coming First World Debt Crisis* (Basingstoke, UK: Palgrave, 2006).

21 The examples in this paragraph are all from Sejerstad, *Age of Social Democracy*, 152–54.

22 See, for example, Berman, "Social Democracy."

23 Hannah Arendt, *Origins of Totalitarianism* (1973; New York: Harvest Book, 1976), 296.

24 Esping-Andersen, *The Three Worlds of Welfare Capitalism* (Cambridge: Polity Press, 1989), 12, my emphasis.

25 Richard Sandbrook, Marc Edelman, Patrick Heller, and Judith Teichmark, *Social Democracy in the Global Periphery: Origins, Challenges, Prospects* (Cambridge: Cambridge University Press, 2007), 127.

26 John Podesta and Casey Dunning, "We Can End Poverty, but the Methods Might Surprise You," *Comment Is Free* (blog), May 7, 2013, http://www.theguardian.com/commentisfree/2013/may/07/end-poverty-use-social-safety-nets.

27 Nihal Bayraktar and Blanca Moreno-Dodson, "How Can Public Spending Help You Grow? An Empirical Analysis for Developing Countries," *Bulletin of Economic Research* 67, no. 1 (2015): 30–64.

28 A similar, more technical definition can be found in Huck-Ju Kwon, Thandika Mkandawire, and Joakim Palme, "Introduction: Social Policy and Economic Development in Late Industrialisers," *International Journal of Social Welfare* (2009): S1. To be clear, I am not seeking to transfer a given view of "society" or "social forms" to other places here, to gift them like so many bags of grain. I am seeking rather to work politically upon what the social is taken and given to mean: a task that in many cases amounts to creating new spaces for social realisation outside of the grip of capital. Politics cannot itself ever be separate to the social (though it arises prior to it). But it can seek to rework the ground upon which we stand, to present and promote

other ideas of what society is than simply those that emerged (and we have largely adopted) along with the rise of capitalism. Indeed, it is supposed to be more than simply where we are at and what we presently think about this: politics is not administration or police; that is to say, it is thought that leads to action in advance of what is possible.

29 Thandika Mkandawire, "From the National to the Social Question," *Transformation: Critical Perspectives on Southern Africa* 69 (2009): 130–60.

30 Esping-Andersen, *Three Worlds*, 14.

31 Kwon, Mkandawire, and Palme, "Introduction," S6.

32 Ibid.

33 See Jimi O. Adesina, "In Search of Inclusive Development: Social Policy in Sub-Sahara Africa Context," in *Social Policy in Sub-Saharan African Context: In Search of Inclusive Development* (Basingstoke, UK: Palgrave, 2007).

34 See Mkandawire, "From the National to the Social Question."

35 Peter Eigen, "A Prosperous Africa Benefits Everybody," *Journal of World Energy Law and Business* 7, no. 1 (2014): 6.

36 Manuel Riesco, *Latin America: A New Developmental Welfare State Model in the Making?* (Basingstoke, UK: Palgrave Macmillan and UNRISD, 2007).

37 See Lena Lavinas, "21st Century Welfare," *New Left Review*, November–December 2013, 30.

38 Data from the *Financial Times*, at http://www.ft.com/intl/cms/s/0/383530cc-8afe-11e2-b1a4-00144feabdc0.html#axzz2VvICDvMO.

39 On the Brazilian experience and the Partido dos Trabalhadores, see Gianpaolo Baiocchi, Einar Braathen, and Ana Claudia Teixeira, "Transformation Institutionalized? Making Sense of Participatory Democracy in the Lula Era," in Stokke and Törnquist, *Democratization*, 231. See also the sage comments on state (*institutionalista*) versus civil society (*participatista*) perspectives outlined by Patrick Heller in the same volume: "Participation and Democratic Transformation: Building Effective Citizenship in Brazil, India and South Africa," 42–74, 49–53.

40 For background on Brazil, see Moses N. Kiggundu, "Anti-Poverty and Progressive Social Change in Brazil: Lessons for Other Emerging Economies," *International Review of Administrative Sciences* 78 (2012): 733.

41 Sandbrook et al., *Social Democracy*, 102–6.

42 Ibid.

43 Ibid., 62–69.

44 Gunnar Myrdal, "The Need for Reforms in Underdeveloped Countries," in *The World Economic Order: Past and Prospects*, ed. Sven Grassman and Erik Lundberg (New York: St. Martin's Press, 1981), 501–25.

45 See, for example, Björn Dressel, "Targeting the Public Purse: Advocacy Coalitions and Public Finance in the Philippines," *Administration & Society* 44,

no. 6 (2012): 65S–84S. There is indeed a range of welfarist approaches, many of which do not resemble the northern models at all. See Rebecca Surender and Robert Walker, eds., *Social Policy in a Developing World* (Cheltenham, UK: Edward Elgar, 2013).

46 Cf. Sandbrook et al., *Social Democracy*, 33.

47 Heather Stewart, "£13tn Hoard Hidden from Taxman by Global Elite," *The Guardian*, July 21, 2012.

48 On the UK tax system and the examples I cite, see Polly Toynbee, "This Farcical Tax System Is Cheating Us out of Billions," *Comment Is Free* (blog), July 29, 2014, http://www.theguardian.com/commentisfree/2014/jul/29/farcical-tax-system-cheating-billions-chase-avoiders. See also ActionAid, *Collateral Damage* (March 2012), http://www.actionaid.org.uk/sites/default/files/doc_lib/collateral_damage.pdf, 1.

49 See Eigen, "Prosperous Africa," 7.

50 Bretton Woods Project and Latindadd, "Breaking the Mould: How Latin America Is Coping with Volatile Capital Flows," December 15, 2011, http://www.brettonwoodsproject.org/art-569425.

51 On the difference, see Bretton Woods Project, "IMF and Capital Flows: All Talk and No Solution," February 7, 2012, http://www.brettonwoodsproject.org/art-569565.

52 Kevin Gallagher, "Cross Border Financial Regulation Is Justified Now More Than Ever," *Global Development* (blog), April 19, 2012, http://www.theguardian.com/global-development/poverty-matters/2012/apr/19/cross-border-financial-regulation-justified.

53 Ibid.

54 Bretton Woods Project, "Time for a New Consensus: Regulating Financial Flows for Stability and Development," December 15, 2011, http://www.brettonwoodsproject.org/art-569411.

55 Jubilee Debt Campaign, *The State of Debt: Putting an End to Thirty Years of Crisis* (May 20, 2012), 29.

56 Ibid., 2.

57 I have found Michael McLure's work useful for thinking with Pigou, and the helpful (if anachronistic) descriptors "distributional fairness" and "macroeconomic stability" are from his "One Hundred Years from Today: A. C. Pigou's Wealth and Welfare" (Discussion Paper No. 12.06, University of Western Australia), http://www.business.uwa.edu.au/__data/assets/pdf_file/0006/2068629/12-06-2012-History-of-Economics-Society-Conference-One-Hundred-Years-From-Today-AC-Pigous-Wealth-and-Welfare.pdf, 3. On global public economics, see especially Tony Atkinson, *Public Economics in an Age of Austerity (The Graz Schumpeter Lectures)* (London: Routledge, 2014), chap. 4.

Press, 2000). For more general perspectives, see Ian Shapiro and Casiano Hacker-Cordon, eds., *Democracy's Edges* (Cambridge: Cambridge University Press, 1999).

83 Ronald Dworkin, "Do Liberty and Equality Conflict?" in *Living as Equals*, ed. Paul Barker (Oxford: Oxford University Press, 1996), 39–57.

CHAPTER 6

1 Carl Schmitt, *The Concept of the Political*, trans. George Schwab (1932; Chicago: University of Chicago Press, 2007), 54 (Schmitt is in dialogue with Proudhon); Adam Smith, *The Theory of Moral Sentiments* (1759; London: Penguin, 2010), 157.

2 For all their presumed similarities, for Smith the governing virtue that should determine the limits to action was always justice; for Friedman it was non-coercion. See Harvey S. James Jr. and Farhad Rassekh, "Smith, Friedman and Self-Interest in Ethical Society," *Business Ethics Quarterly* 10, no. 3 (2000): 659–74.

3 Istvan Hont, *The Jealousy of Trade: International Competition and the Nation State in Comparative Perspective* (Cambridge, MA: Harvard University Press, 2010).

4 The term is Primo Levi's. See Juan Manuel Iranzo, "Trilogía de Primo Levi," *Revista de Libros*, no. 108, December 2005, http://www.revistadelibros.com/articulo_imprimible.php?art=3259&t=articulos.

5 The phrase is given a much fuller exegesis in Claude Lefort, *The Political Forms of Modern Society: Bureaucracy, Democracy, Totalitarianism* (Cambridge, MA: MIT Press, 1986).

6 See, for example, Anne-Emmanuelle Birn, *Marriage of Convenience: Rockefeller International Health and Revolutionary Mexico*, Rochester Studies in Medical History (Rochester, NY: University of Rochester Press, 2012).

7 On Trump, see "Donald Trump: Still a Miserly Billionaire," *Smoking Gun*, January 17, 2012, http://www.thesmokinggun.com/documents/miserly-donald-trump-654712.

8 David McCoy, "Conflicts of Interest within Philanthrocapitalism," *Global Health Watch III* (London: Zed Books, 2011), 267–74.

9 Amartya Sen, *The Idea of Justice* (London: Allen Lane, 2010).

10 See I. M. Young, "From Guilt to Solidarity: Sweatshops and Political Responsibility," *Dissent*, Spring 2003, 39–45; Young, "Responsibility and Global Labor Justice," *Journal of Political Philosophy* 12, no. 4 (2004): 365–88; and Young, "Responsibility and Global Justice: A Social Connection Model," *Social Philosophy and Policy Foundation* 23, no. 1 (2006): 102–30.

11 Nancy Fraser, "Can Society Be Commodities All the Way down? Polanyian Reflections on Capitalist Crisis" (FMSH-WP-2012-18, 2012), 10.

12 Andrew Sayer, *Why We Can't Afford the Rich* (Bristol, UK: Policy Press, 2015), 94.

13 Ruth Marcus, "Moral Dilemmas and Consistency," *Journal of Philosophy* 77, no. 3 (March 1980): 121–36; see also Jessica B. Payson, "Moral Dilemmas and Collective Responsibilities," *Essays in Philosophy* 10, no. 2 (2009): art. 4.

14 Andrew Schaap, "Guilty Subjects and Political Responsibility: Arendt, Jaspers and the Resonance of the 'German Question' in Politics of Reconciliation," *Political Studies* 49, no. 4 (2001): 749–66. See also Doreen Massey, "Space, Time and Political Responsibility in the Midst of Global Inequality," Erdkunde 60: 89–95, 94.

15 The first quote is from Christopher Kutz, *Complicity: Ethics and Law for a Collective Age* (Cambridge: Cambridge University Press, 2000), 1. The second is from John Gardner's "Review: Christopher Kutz, Complicity: Ethics and Law for a Collective Age," *Ethics* 114, no. 4 (2004): 827–30, 828.

16 This is not to say we can afford to do away with these; we need both.

17 Cited in Ira Howerth, "The Social Question of Today," *American Journal of Sociology* 12, no. 2 (1906): 254.

18 Robin Blackburn, "Reclaiming Human Rights," *New Left Review*, May–June 2011, 126–38.

19 Adam Smith, *The Wealth of Nations* (New York: Random House, 1937), 681.

20 Onora O'Neill, "What We Don't Understand about Trust," Ted Talk, http://www.ted.com/talks/onora_o_neill_what_we_don_t_understand_about_trust.

21 For a discussion of this, see Gerry Kearns, "Progressive Geopolitics," *Geography Compass*, 2 no. 5 (2008): 1599–1620, 1606.

22 Michael Edwards, "When Is Civil Society a Force for Social Transformation," *OpenDemocracy*, May 30, 2014, http://www.opendemocracy.net/transformation/michael-edwards/when-is-civil-society-force-for-social-transformation.

23 Gøsta Esping-Andersen, *Politics against Markets: The Social Democratic Road to Power* (Princeton, NJ: Princeton University Press, 1988).

24 There are many examples of this, including the re-framing of Costa Rica's state-led development achievements of the 1950s to 1970s as a case of neoliberal "success" in the 1980s and 1990s. Richard Sandbrook, Marc Edelman, Patrick Heller, and Judith Teichmark, *Social Democracy in the Global Periphery: Origins, Challenges, Prospects* (Cambridge: Cambridge University Press, 2007), 110.

25 Francis Wilcox, foreword to Gunnar Myrdal, *The Challenge of World Poverty: A World Anti-Poverty Programme in Outline* (New York: Pantheon Books, 1970), vi.

ACKNOWLEDGMENTS

This book was conceived as a short essay but evolved, at Christie Henry's editorial encouragement, into the larger argument you have before you. Inequality matters to all of us, and we all have something to gain by taking active measures to address it. The purpose of these pages has been to make the case both for why this is and for why we should not confine this discussion to the rich world alone.

Writing the book would have been impossible without the award of a Philip Leverhulme Prize in 2011, and I am extremely grateful to the Leverhulme Trust for supporting my work. The book was also shaped by an informative period attached to Columbia University's Department of History and the Heyman Center for the Humanities in New York in 2013, funded as part of the Leverhulme programme of work. I thank in particular Mark Mazower and Eileen Gillooly for facilitating this. I am grateful as well to Fritt Ord and the Norwegian Authors' Union for their support and to the staff of Oslo's Litteraturhuset for providing me with a place to write. Like much else in my life of late, a good few parts of the book were written somewhere in the air between London and Oslo.

Whatever the book's flaws, inevitable in so wide-ranging a disquisition, these are mine alone to answer for. Such worthwhile contributions as the book makes, on the other hand, have likely benefitted from the conversations and support I have received from numerous individuals, among whom I must mention in particular Jonathan Glennie, Håvard Friis Nilsen, Kristian Stokke, Kristoffer Lidén, Magnus Marsdal, Samuel Moyn, Partha Chatterjee, Ole Jacob Sending, and Stuart Corbridge, in whose classes on development studies at Cambridge University in the 1990s the political origins of my own interest in this topic are to be found. The final manuscript has benefitted from generous yet critical readings by Alan Lester, Danny Dorling, Alec Murphy, Jens Plahte, Ola Innset, Adrian Smith, Jane Wills, David Nally, Emma Mawdsley, Matt Sparke, Elizabeth Day, Alistair Taylor, and two anonymous referees. At University of Chicago Press I have again been hugely privileged to work with Christie Henry and her wonderful editorial team: Abby Collier, Amy Krynak, Katherine Faydash, and Melinda Kennedy. At Capel & Land I must again thank George Capel for helping bring the book to fruition.

I am grateful too for the assistance extended me in a range of other ways. Nicole Lieberman provided additional research support, efficiently tracking down and compiling obscure references during a period when work-life balance was a circle that could not go on being squared. Lars Gunnesdal produced the graphs in chapter 1. Ed Oliver redrew the map that is also found there, amended it innumerable times, and then promptly drew it all again from scratch at short call. Ed is testament to the intense collegiality with which the School of Geography at Queen Mary, University of London—my intellectual home the past ten years—is infused. Yet again I must express my gratitude at being part of the School of Geography's wonderful community of scholars. I am no less grateful for the equally collegial home provided over the past few years by staff and colleagues at the Peace Research Institute Oslo. There I am especially indebted, for their friendship as well as their professional insight, to Kristoffer Lidén, Kristin Sandvik, Pinar Tank, Maria Gabrielsen, Kristian Hoelscher, and Jason Miklian.

Collegiality comes in many different forms, however, and some of the best of them are Mediterranean. For this I am grateful to Ritsa Balatsoura for a place to work in Pelion while finishing a first draft of the book, and to Ritsa Storeng, my mother-in-law, for her regular Sunday dinner table in Oslo, around which a growing family congregates. There it has been possible to take leave of the problems of an unequal world amidst the inventive inclusions of a Greek-Norwegian-British family. Like its author, this book owes most of all to my wonderful and brilliant wife, Katerini. It is written for our children, in the hope that they might one day know what is to be done. But it is dedicated to the memory of my father, whose books on the classics of political economy I first borrowed long ago, in search of answers myself.

FIGURE CREDITS

Figure 1. Here, as elsewhere in this chapter, I have redrawn and updated some "classic" figures. The original source for this graph (except for Ghana) is Angus Maddison, *Statistics on World Population, GDP and Per Capita GDP, 1-2008 AD*, OECD Development Centre (March 2010), http://www.ggdc.net/maddison/oriindex.htm. All data used to redraw it come from J. Bolt and J. L. van Zanden, *The First Update of the Maddison Project: Re-estimating Growth before 1820* (Working Paper No. 4, Maddison Project, 2013). Data are available for download at http://www.ggdc.net/maddison/maddison-project/home.htm.

Figure 2. Branko Milanovic, *Worlds Apart: Measuring International and Global Inequality* (Princeton, NJ: Princeton University Press, 2005). Information for years 1990–2010 has been updated using calculations from the Conference Board of Canada: "The 1990 to 2010 results were calculated by The Conference Board of Canada using World Bank data for 172 countries for which data on per capita income were available. These 172 countries account for 98 per cent of world gross domestic product and 97 per cent of world population. The Conference Board data were adjusted to account for differences in country coverage and data between Milanovic's study and that of the Conference Board." The Conference Board data are available at "World Income Inequality," Conference Board of Canada, http://www.conferenceboard.ca/hcp/hot-topics/worldinequality.aspx.

Figure 3. *Panels A–C*: Data for Gini (latest year), child mortality rates (2012), Education Index (2013), and Health Index (2013) from UN Human Development Reports 2014. Additional Gini data from OECD and Eurostat. Data available at http://hdr.undp.org/en/data. *Panel D*: K. Deininger and P. Olinto, "Asset Distribution, Inequality, and Growth" (Policy Research Working Paper No. 2375, World Bank), 24, table 2, http://papers.ssrn.com/sol3/papers.cfm?abstract_id=630745.

Figure 4. Branko Milanovic, *Worlds Apart: Measuring International and Global Inequality* (Princeton, NJ: Princeton University Press, 2005).

Figure 5. Branko Milanovic, "Global Income Inequality," IMF Institute, March 2010, http://siteresources.worldbank.org.

Figure 6. The original map is from TD Architects and is available at the website Information Is Beautiful (http://www.informationisbeautiful.net). There are a number of flaws in the original map, however, so I have redrawn it here to correct for them. I have added the visualisation of the number of deaths of irregular migrants seeking to enter wealthier lands at these same borderlands. The map is still far from perfect, and it should be read with caution. Not least (as with any map) the selection criteria are arbitrary. But it remains, all the same, a not-unrepresentative visualization of the issue.

Figure 7. *Panel A*: J. K. Galbraith, "Global Inequality and Global Macroeconomics," *Journal of Policy Modeling* 29 (2007): 587–607, http://www.ideaswebsite.org/pdfs/global_inequality.pdf. *Panel B*: Data from the University of Texas Inequality Project–UN Industrial Development Organisation (UTIP-UNIDO) database. Data available for download from http://utip.gov.utexas.edu/data.html.

Figure 8. Kate Wilkinson and Richard Pickett, *The Spirit Level: Why More Equal Societies Almost Always Do Better* (London: Allen Lane, 2009), 61.

Figure 11. Sergey Ponomarev/The New York Times/Redux.